Praise for *Evolution 2.0*

"Standing on the knife-edge between traditional evolutionary theory and Intelligent Design, this book will inflame both dogmatic Darwinists and Creationists. It's irritating to both because it's friendly to the idea of evolution itself and because it judges Darwinism too close-minded and reductive.

In writing *Evolution 2.0*, my friend Perry Marshall has chosen the path of maximum risk, but with it the chance of a pioneering new horizons in the origin and evolution of life. Perry has deliberately parachuted into hostile territory. To bridge this gap demands rare qualities: a maverick approach, thirst for challenge, freedom to explore, and the will to slaughter sacred cows.

Perry dares to bring new disciplines to the debate, namely computer science and electrical engineering. These fields bring light and innovative problem-solving to biology.

While most scientists submit to self-censorship and dare not question cherished assumptions, the boldest scientists like Albert Einstein, Francis Crick, and Stephen Hawking never feared such taboos. Whether agnostics, atheists, or believers, they never shrank back from big questions or unconventional solutions. Evolution and design are not either/or, but both/and."

—JEAN-CLAUDE PEREZ, author, *Codex Biogenesis*, and retired IBM Biomathematics and Artificial Intelligence Interdisciplinary Researcher

"In *Evolution 2.0*, Mr. Marshall has made a gallant attempt to bridge the gap between conventional evolutionary theory and creationism by applying his vast knowledge in computer science and electrical engineering to biology.

Mr. Marshall is making an invaluable contribution toward more open and honest discussion on the subject of evolution versus creation. The book is well written, often witty, and is extremely thought provoking.

I pre-ordered a few copies of *Evolution 2.0* for our grandchildren. It is amazing that this 'non-biologist' has analyzed life phenomena at the cellular and molecular levels to such depth and width by applying information gained by cutting-edge methods in the life sciences!

The author has amassed a wealth of information gathered from 409 cited sources in various fields, ranging from biology to theology. Being a cell/molecular biologist and a Christian, I have often been asked about my views on Christian faith and evolution by my students and colleagues. I have responded by saying I do not see any conflict between one's personal belief in creation and scientific search for the mechanism of evolution.

This is because I believe the recognition of God as the creator and sustainer of life does not have to be in opposition to the concept of evolution, more complex organisms arising from simpler ones. However, I recognize that the conflict between the atheistic view of human origin as a result of mechanistic evolution and that of human creation in God's image might remain unresolvable.

Mr. Marshall offers a technology prize to anyone finding 'an example of information that does not come from a designer.' The prize money alone would be an incentive to read the book, especially for those who deny the existence of a designer."

—Dr. Kwang Jeon, editor, *International Review of Cell and Molecular Biology*, and professor emeritus, Department of Biochemistry, University of Tennessee

"Any person of faith who cares about how creation reveals God—and how evolution is actually a devastating arsenal of evidence *against* atheism—should add *Evolution 2.0* to the extreme tippy-top of their reading list. *Evolution 2.0* is equally shocking to both atheists and Christians, a genuine eye-opener in a category of its own.

Perry has a dogged focus in the finest tradition of science's founders. He's fair and honest with facts that few assess calmly. He explains complex evolutionary systems with clear and understandable illustrations. He's superbly crafted an engaging and persuasive narrative.

Evolution 2.0 weaves seemingly dry, technical, even incomprehensible topics into a tight, fascinating story around his own scientific and spiritual journey—revealing empirically valid and truly astonishing facts about DNA. These facts turn mainstream evolution on its head.

He demonstrates beyond doubt that the development of new attributes and species is staggeringly sophisticated, directly comparable with computer code and languages. . .only far more advanced. All of this is extensively referenced for readers who want to verify his claims or just discover amazing capabilities of DNA that have been proven (but not publicized) for 50+ years.

Perry's focus on scientific data does not come at the expense of other critical issues. He ably canvasses the hard problem of consciousness, the philosophical underpinnings of science, the curiosity-killing presuppositions of neo-Darwinism, and science's history with faith. He is commendably straightforward and presents the central ideas clearly—even to people with no prior knowledge.

He may or may not convince you of common ancestry or the age of the earth. But you will find his overall presentation absolutely compelling."

—D. BNONN TENANT, ThinkingMatters.org.nz

"A very readable book and a devastating attack on the neo-Darwinist orthodoxy that evolution is nothing but natural selection acting on random variation."

—PETER SAUNDERS, codirector, Institute of Science in Society, and Emeritus professor of Applied Mathematics, King's College, London

"With considerable wit and amazing insight, Marshall delivers a compelling and forceful synthesis that sets a new standard for discussions about the relationship between science and faith. The result is no less than astonishing."

—MARK MCMENAMIN, professor of geology, Mount Holyoke College

"Perry shows that there are processes in the cell which suggest that the cell itself is sufficiently smart to rearrange its genome and direct its own evolution. Perry argues that at some level there may be a relationship between naturalistic processes and design reminiscent of the paradox of particle/wave duality in physics.

The book is wonderfully thought provoking as Perry brings a fresh perspective to the increasingly arid (and acrimonious) debate between Intelligent Design and methodological naturalism."

—CROFTON BRIERLEY, MS, Biochemistry, Oxford University, former department head, Marconi Optical

"A remarkable and useful resume of the state of the art of this great problem of science."

—STUART PIVAR, author, *Lifecode* and *On the Origin of Form*, and cofounder, New York Academy of Art

"I am committed to Young Earth Creationism, and thus cannot agree with some of Perry's conclusions. However, I highly recommend this book to any skeptic who is committed to a purely materialistic paradigm. The science presented here—from the latest research to the most engaging minds on this subject—makes this the one book you should read."

—RAY GLINSKI, MS, Biochemistry, pastor at Grace Church of DuPage

"*Evolution 2.0* is a modern philosophical marvel unlike anything I have read in my years of study. It allowed me to put down my guard. The author was not compelling me to believe in an ideology, but rather taking me alongside his journey of self-discovery. I came into this book hesitant. As a staunch creationist I found myself trying to fight with Marshall early on, but his arguments and presentation were flawless and compelling. It was refreshing to see someone examine both sides honestly.

While reading his provocative and thorough discourse, I was not offended with the manner in which he dealt with a sensitive topic. Marshall was raw and candid, but curious. He was never hasty to draw conclusions, but he also didn't drag the reader through an endless chamber of mind maps. Marshall's approach was logical, while retaining a human element. He takes us through his journey, and in turn, we enter our own voyage of discovery. Iron sharpens iron.

Evolution 2.0 treats both science and theology with respect that is merged with a healthy skepticism. Given the complexity of the topic, coupled with Marshall's expertise in the area, this logical process is necessary.

Evolution 2.0 caters to a scientific mind, but can be appreciated by all. It is well written and orderly. This book explores a very complex and heavy matter; but at the same time, the author interjects a very personal tone all while maintaining empirical integrity."

—Marie Sarantakis, comparative religion scholar, Carthage College

"As one maverick critiquing another, I admire *Evolution 2.0* by Perry Marshall. Armed with computer science and electrical engineering, Perry fights an uphill battle to unite the space between those who believe evolution is random and those who believe species are designed by God, who in some cases deny evolution itself.

Some will never yield their 'God-given right to be atheists.' For them, Perry's fluid reasoning, his vivid, readable explanations, easy style, and enjoyable storytelling may be deemed 'unreasonable' or 'argued to death.'

Unless, of course, someone wins the technology prize! Should that happen, nobody will argue with success. Until then, people will be debating this book for years.

Judge this book by the science within its pages—and enjoy the story."

—Andras Pellionisz, biophysicist, founder of Fractogene; PhD, computer technology, PhD, biology, PhD, physics

"Sometimes the only way a puzzle gets solved is when someone looks at it with fresh eyes. Perry Marshall is an engineer who started to investigate biology. His book could signal a paradigm shift in the battle between Darwinian evolution and creation/ID. Maybe the war is over. Respond, criticize, and debate it...just don't dismiss it."

—JUSTIN BRIERLEY, host, *Unbelievable?* radio program and podcast

"Perry's book changed my life forever. Here was a guy who was offering answers to questions I didn't even know I had. Things that bugged me for so long suddenly made sense.

When I read Perry's book, my world tilted. Not only did I learn stuff nobody's talking about, I realized my God was way bigger than I ever gave Him credit for. Long-held, preconceived notions about science, life, and the Bible were shattered. Perry taught me never to fear the truth, and for that I'll always be grateful."

—JETHRO FRANK, age 19, East Chain, Minnesota

EVOLUTION 2.0

Breaking the Deadlock Between Darwin and Design

Perry Marshall

BENBELLA BOOKS, INC.
DALLAS, TEXAS

Further image credits and permissions on page 359.

BenBella Books, Inc.
10300 N. Central Expressway
Suite #530
Dallas, TX 75231
www.benbellabooks.com
Send feedback to feedback@benbellabooks.com

Printed in the United States of America
10 9 8 7 6 5 4 3 2 1

Library of Congress Cataloging-in-Publication Data

Marshall, Perry S.
Evolution 2.0 : breaking the deadlock between Darwin and design / Perry Marshall.
pages cm
Includes bibliographical references and index.
ISBN 978-1-940363-80-6 (trade cloth : alk. paper) — ISBN 978-1-940363-90-5 (electronic) 1. Religion and science. 2. Evolution (Biology) 3. Intelligent design (Teleology) 4. Creationism. 5. Evolution (Biology)—Religious aspects—Christianity. I. Title.
BL263.M4175 2015
213—dc23

2015005090

Editing by Leah Wilson, Heather Angus Lee, and Richard Morgan
Copyediting by James Fraleigh
Proofreading by Brittany Dowdle and Lisa Story
Indexing by WordCo Indexing Services, Inc.
Text design and composition by John Reinhardt Book Design
Cover design by 99designs/W. Antoneta and Sarah Dombrowsky
Printed by Lake Book Manufacturing

Distributed by Perseus Distribution
www.perseusdistribution.com

To place orders through Perseus Distribution:
Tel: (800) 343-4499
Fax: (800) 351-5073

E-mail: orderentry@perseusbooks.com
Significant discounts for bulk sales are available.
Please contact Glenn Yeffeth at glenn@benbellabooks.com or (214) 750-3628.

For all the atheists, agnostics, and skeptics
who spent untold late nights helping me hone
and polish the content of this book:
To you, I am eternally grateful.

Contents

Introduction

Part I

Evolving My Religion

Part II

The Neo-Darwinist Dilemma

Part III

How Evolution *Really* Works

Part IV

Evolution 2.0 and the Language of Cells

Part V

Origin of Information: The Quest

Part VI

Evolution 2.0 and Its Implications for Science

Part VII

Evolution 2.0 and Its Implications for Faith

Conclusion

You've Reached the End of My Story. You Still Might Want to Read Further. Here's Why.

INTRODUCTION

The Young Earth Creationist and His Curious Daughter

MY FRIEND BOB knows I've been obsessing about evolution. He's a Creationist who believes the Earth is 6,000 years old. He doesn't buy into the idea of evolution, but he does think our conversations are entertaining.

One night, Bob, his kids, and I went to a restaurant for dinner. His 12-year-old daughter Melanie was sitting across the table from me. We were talking about evolution and the pros and cons of antibiotics. Her brother Jack was obsessing about the Chicago Bears. Bob was sufficiently secure in his manhood to let me try to convince his daughter that evolution is possible. He also knows I would never attempt to snuff out anyone's sense of wonder.

So, says I, "Hey Melanie, you know about antibiotics, and how you have to be careful with them because germs develop resistance, right?"

"Oh yeah," she says. "If you don't finish your whole prescription, then you *almost* kill the bugs but not quite. Then bugs become superbugs. Infection spreads all over the place and you're dead."

Smart girl.

"Has anyone ever told you *how* the bugs become superbugs?" I asked.

"No."

"Let's say you've got strep throat and you're taking antibiotics," I reply. "The antibiotic poisons the bacteria in your system. It's breaking down their cell walls and it's killing them. So they're swimming around inside you and they go on red alert. It's as though they're saying, 'Hey, this poison is killing us! We've got to find a way get this poison out!'

"They troll through your body searching for a solution. Eventually, one finds a cell that has a pump that could pump the poison out. That cell offers the bacterium the DNA in its plasmid. A plasmid is a file-sharing folder for swapping DNA. The bacterium pulls that DNA inside its cell wall. It finds the section of DNA that codes for a pump, reads the code, and builds itself a pump."

Melanie blinks hard, listening intently.

"Then if the pump works, the cell divides in two and makes daughter bacteria cells, and they inherit the pump. Not only that, it becomes a little software salesman. It starts giving code for its new pump to all its bacteria friends."

Melanie furrows her brow.

"That's why antibiotics eventually stop working. But your own immune system fights back in similar ways. Immunity is actually an arms race between your body's cells and the invading cells."

Melanie stops mid-bite, eyes wide with wonder. "Whoa... that's *cool*!"

I turn to Bob. "That is just a taste of how evolution *really* works. Bugs become superbugs in one generation."

Bob is chewing his last bite of steak. "I didn't know that, and maybe it's all true," he says, "but evolution still gives me the heebie-jeebies. Okay, God *might* have used some evolution to make things the way they are. Things can evolve. But not everything evolved. Especially not by accident."

"Bob, *none* of this happens by accident. You know how you have to get a tetanus shot every 10 years? That's because when you get it the first time, your immune cells add new equipment and instructions. A decade later they realize that they've been lugging it around so long without using it that they don't need it anymore. They discard the excess baggage. The shot shocks your system and makes them say, 'Okay, wait a minute, maybe we *do* need this code.' Your immune system responds by beefing up its defenses again."

Bob's eyes narrow a little. "You make it sound like cells can think for themselves."

I smile. "They even talk to each other! Bacteria live in colonies and greet each other when they meet. They have words for *me*, *you*, *us*, and *them*. Different species speak different dialects. They're second only to humans in linguistic ability."

Bob pushes back: "Yeah, but bacteria fighting antibiotics isn't the same as camels evolving into giraffes. Sure, of course I believe in micro-evolution, but they've never proven macroevolution."

"Macroevolution—quantum adaptive leaps that produce brand new species—gave us wheat, which is a hybrid of two weeds. Somebody came up with it 11,000 years ago. Breeders create new species every day. They cross two different species together to make a third species, literally overnight. It doesn't take millions of years. Then, just like the bacteria rearrange their DNA, the cells of the new hybrid go to work and tune the fine details."

Bob and I have been buddies for years, and he's giving me a familiar look. That look means, *Perry, I like the way you make me think, but some of the stuff you come up with is just strange.*

Melanie takes a sip of Coke and leans toward me across the dinner table. Even though I've been mostly talking to Bob for the last few minutes, she's tracking perfectly.

"Cells actively respond to enemies and threats?" she asks. "And borrow DNA from other cells? And when you mix two different species together, you get a new species?"

"Yeah," I reply. "Most of the time the new species is sterile. But sometimes it's not. And the funny thing is, the jumps happen *fast*. Weeds to wheat didn't take millions of years, it might have only taken 100 generations to reach its current form."

Bob's über-conservative. He's not entirely convinced, but he's intrigued. "So all this could all happen in less than 10,000 years?"

"It's not that fast. The Earth is very old; the high school biology books got that right. But the average biology book tells you that copying errors of DNA—'random mutations'—create gradual changes over time that are then filtered by natural selection to create new species, and the latest research shows this is wrong. Bacteria, plants, and giraffes adapt by exchanging and rearranging their DNA. Cells are tiny programmers. They rewrite their code when they're under stress."

Bob smiles a little as if I've allowed him a small victory. "So it's not random at all."

"Nope. It's not random; it's engineered. It's not gradual; it's quantum leaps. Cells adapt. The results they achieve are far superior to our own. There's a lot we can learn by studying cells. And evolution."

The waitress brings the check. Melanie's brother, Jack, is still talking about the Bears' upcoming match against the Green Bay Packers, so we switch to that. I notice that Melanie is lost in thought—she hasn't swallowed her next sip of Coke yet.

I think, *if I'd known what Melanie just discovered when I was 12, I might have become a biologist instead of an engineer.*

The Road to Code

FOR MOST OF MY LIFE, I suspected evolution might be a fraud. As commonly depicted, it felt like an insult to engineers, artists, and all hard-working creative people. But then an argument with my brother (which you'll soon hear about) forced me to look deeper and question all my assumptions. I could not imagine how much more amazing the world was about to become, and my discoveries only served to confirm: *Truth really is stranger than fiction.*

One day I had a huge epiphany: I suddenly saw the striking similarity between DNA and computer software. This started a 10-year journey that led me down a long and winding road of research, controversy, and personal distress before I discovered a radical re-invention of evolution.

DNA is code. And before you're even halfway through my story, you'll discover that this one simple fact, firmly established in the 1960s, not only forms the very bedrock of modern genetics but holds sweeping implications for science, technology, even religion.

In the 21st century, we know as much about codes as we know about almost anything else in science. Everyone who has a computer or sends text messages on their cell phone uses codes. Every major university in the world has a computer science department where you can take undergraduate and graduate programming classes and learn to design codes. We pay educated people handsome salaries to develop even better codes and help us store, process, and transmit data.

We understand codes better than we understand gravity, the laws of thermodynamics, or quantum physics. No one knows how to create gravity, but millions of people know how to create code, including some of the wealthiest businesspeople in the world.

Everything we know from computer science provokes a huge question: *How do you get a code without a coder?* And, as you'll soon start asking: *How can code write itself?*

These questions challenge the boundaries of science and religion.

Show Me the Science

Secular science has no answer to these questions. They refuse to go away. Conversations about faith can bring these questions into sharp focus. Yet our mass media pits science against faith, and vice versa, as if the two are by necessity mutually exclusive.

Not everyone believes this to be the case. In just a few pages we'll dive into these questions, but first it's important to understand the spectrum of beliefs surrounding evolution and creation.

Evolution has come quite far since Charles Darwin first rolled out his theories in 1859. Nevertheless, the matter of "Where do we come from?" and "How did we get here?" is debated as hotly today as ever.

Who are the most vocal players in this debate? Meet the extremes in the creation-evolution debate.

> **(Young Earth) Creationist:** Person who believes the universe was created by God in six literal 24-hour days.*

With its insistence on a series of creative miracles, Creationism doesn't give empirical science a chance. Creationism rejects modern dating methods and large portions of geology, paleontology, and astronomy with the belief that the entire universe was made by God in six literal 24-hour days.

In many circles, Creationism has morphed into the position of Intelligent Design, or ID.† ID, while recognizing many truths about biology

---

* Dictionary.com, s.v. "creationism," accessed January 15, 2015.

† Intelligent Design has been accused of being nothing more than "Creationism in a cheap suit." But it's important to note that for many ID advocates, God has little to do with ID. There's an important distinction between IDers who believe in episodes of divine intervention and IDers who, often apart from religion, observe that mindless, materialistic processes simply fail to explain or adequately describe many aspects of living things (see Discovery Institute at http://discovery.org/about, accessed January 13, 2015). They may not have an answer for the origin of the design, and they may or may not think it's divine, but to them the question is secondary to the task of science itself. ID asserts that the same principles of design employed in architecture,

that old-school Darwinism denies, ultimately abdicates its responsibility by jumping directly to "God did it." At least in its most simple forms, ID halts scientific inquiry by dismissing too easily the possibility that God may have used a *process* to develop life on Earth. Further investigation becomes impossible if a miraculous event cannot be reproduced in the lab.

A great many biologists reject ID. They believe that genetic variation and natural selection, multiplied over billions and trillions of creatures, produce the *appearance* of design (111). A great many credentialed professionals insist there is no design at all, just random mutations and natural selection.

Now if the secular biologists are right, then the hand at the end of your arm only *appears* to be designed. But it's not designed at all, which means... NO GOD NECESSARY. Many people feel that science has driven God to the fringes, and is only steps away from eliminating religion and spirituality entirely.

Which brings us to the opposite end of the divide—the Neo-Darwinists.

> **Neo-Darwinist**: Person who believes in the "Modern Synthesis" of evolution, building on the idea Charles Darwin expressed in his book *On the Origin of Species* in 1859: that life evolved through a purposeless process of random mutations and natural selection (108, 645).

Hard-line Darwinism downplays the astonishing ingenuity of living things. In the Modern Synthesis, everything happens slowly and without plan. Darwinism invokes a long series of lucky breaks one could scarcely hope to reproduce in the lab. We have to just take someone's word about what really happened. Evolution can seem more like an unverifiable hypothesis and materialistic doctrine than testable scientific theory.

What the extremes have in common is insufficient empirical science—the kind that can be measured, analyzed, and assessed for evidence and accuracy in real time. Meanwhile, there are many ingenious experiments

---

computer science, manufacturing, and music are valid and necessary in science and biology. One need not care about theological questions to recognize that Darwinism fails to answer science questions as well. In the pages to come I'll describe why, from an engineering and technology point of view, ID raises questions we cannot afford to ignore—because they are not only scientifically sound but commercially valuable.

nobody hears about, many thoughtful dialogues between religion and science…and so many shades of gray in the middle, carefully considered views that get lost in the shouting matches between the Creationists and the Neo-Darwinists.

As I distanced myself from the extremes, I found myself reading more scientific papers and fewer popular books and websites. Scientific papers undoubtedly have biases, but at least the authors have to present their data.

That science is what this book is about.

What You Can Expect from This Book

This is not a religious book. If your number-one frame of reference for questions about evolution is Genesis, Young Earth Creation, Old Earth Creation, the Bible, or theology, please start by reading appendix 2 right now—then return and continue. (I've found many of my Christian, Jewish, and Muslim friends need to address specific religious questions first in order to become comfortable with the science that comes in the next few chapters.)

This is a science book, provoked by my burning question: If blind evolutionary forces can produce eyes and hands and ears and millions of species, then why don't engineers use Darwinian evolution to design cars or write software? Why don't they teach Darwinism in engineering school? Evolution and natural selection, after all, were heralded as all-powerful, to the point of having godlike qualities. If nature needs no engineers, a little evolution knowledge would surely be useful to us engineers who are stuck in cubicles designing cell phones.

The answers took me by surprise. Rather than brushing aside "ultimate questions," the new discoveries I was making only served to intensify and reframe ancient questions that lie at the boundary between science and religion. Where did life come from? Where do we get our ability to think and choose?

In this book you'll discover vital principles from the information sciences that neatly explain why, after over 150 years, Darwinism is still plagued with problems, never able to silence its critics—unlike other theories like quantum mechanics and the Big Bang, which have quietly vindicated themselves. You'll discover that living things are more amazing, adaptive, and creative than most people dare to imagine.

You'll also find that the Creationists and ID advocates leave out vital parts of the story: Darwinists underestimate nature, and Creationists underestimate God.

If you care about science (and I think you do, or you wouldn't have picked up this book), then, by the time you're finished reading, you will not be able to accept the explanation of "happy chemical accidents" as the source of life, as put forth by well-publicized atheists (101).

Who Should Read This Book?

If you're a person of faith, and you've been struggling to integrate scientific evidence with your core beliefs, this book is a great start. Not only will it resolve age-old tensions, you'll also literally see every green leaf and blade of grass in a new light.

If you're interested in evolution for evolution's sake, if you simply love science and nature and exploring new things, you'll find much to feed your mind and fuel your imagination.

And if you believe science is a practical endeavor, useful to you in whatever you do; if you're a programmer, an engineer, a medical professional, or business manager; if you're a strategist of any kind in any profession—then this book will reframe what you already do. You'll be inspired and invigorated by the amazing engineering capabilities of the cell.

Who Should *Not* Read This Book?

If you're a staunch six-day Creationist; if you hold a firm conviction that the universe is young and no other interpretation of ancient texts is permissible; if evolution seems an impossible hoax; then you will find this book threatening.

If you're on the opposite end of the spectrum; if you're confident that the major operating principles of the universe are well understood, and science exploration from here on out is just cleanup; if you're certain that reason, logic, and science relegated God to the dustbin decades or centuries ago... then the research I cite in this book will make you squirm, too.

I Put My Money Where My Mouth Is

In this book, I offer a 2.0 version of evolution, a brand new paradigm for biology. I will show you that scientists create new species in the lab every day, and I'll show you how they do it.* I'll also demonstrate that to the extent science can prove anything, science proves design in DNA. In other words, I'll prove that *both sides, the Creationists and the Darwinists, are right*. Yes, I know that's a strong statement. In the coming pages I will back this up with hard science.

This new paradigm is so important that I've organized a private investment group to fund a multi-million dollar technology prize. Similar in some ways to the $10-million XPrize that incentivized the world's first reusable commercial manned spacecraft, I am offering an award to the first person who can discover a process by which nonliving things can create code.

Before life can reproduce and before evolution has any chance of occurring, there must first be a code. Currently we have no evidence to suggest that the genetic code, or any code, can come into existence without intelligence. This prize highlights and seeks to fill a crucial gap in our present understanding of science; it speaks to the Origin of Life question and promises to unlock the secrets of true Artificial Intelligence.

ONLINE SUPPLEMENT

Technology prize as alternative to traditional research funding

www.cosmicfingerprints.com/supplement

The details of this prize are found in chapter 23 and appendix 4. You can check the current status and cash value of the prize at www.naturalcode.org.

If that seems like a bold move, it is. I am a mortgage-holding, taxpaying American guy with a family to support, and I don't make a habit of wildly speculative ventures like this. But the questions this journey raised for me are too urgent to be ignored.

Now, here's my story.

---

* The essence of the term *species* is that members of the same species can interbreed. Two animals are not of the same species if they can't interbreed. In chapters 15 and 16 I discuss mechanisms that produce new species.

PART I

EVOLVING MY RELIGION

CHAPTER 1

"Bro, I'm Losing My Religion"

How many times is the truth that you take to be true
Just truth falling apart at the same speed as you
Until it all comes away in a million degrees
And you're just a few pieces of fallin' debris

—Josh Ritter

THE DIVIDE BETWEEN EVOLUTION and Creation cost my younger brother his faith.

My brother Bryan got his master's degree in theology at a very conservative seminary in California in 1999, but after teaching English in China for a couple of years, his entire belief system unraveled. Before I knew it, he was teetering on the edge of atheism. Questions about Creation and evolution added much fuel to the fire.

During one of my visits, we rode together on a bus in Yunnan province in the foothills of the Himalayas. We wound our way up lush, verdant mountains, looking across vast, contoured valleys where peasants farmed much as they had for thousands of years. There, Bryan confided his concerns to me.

"Perry, I've scoured the internet and there's scarcely a geologist or astronomer anywhere who believes the Earth is 6,000 years old. There are millions of fine layers of sediment in the Earth's crust, deposited year by year."

No problem there, pal. Despite our Young Earth Creationist upbringing, I'd reached the exact same conclusion long before: The Earth is very old. That had not been much of a struggle, because I'd encountered smart people who simply read Genesis a little differently. I was content with their Old Earth views. (See appendix 2.)

But Bryan also posed many other questions for which I lacked answers. As our conversation intensified, I felt a creeping sense of unease. I found myself growing defensive. Eventually I retreated to what I was most comfortable with: science.

I said, "Bryan, I'm an electrical engineer and I've spent 20 years of my life building and designing things, balancing delicate tradeoffs between performance, price, and quality. C'mon dude, the idea that any kind of blind, accidental process can produce the fantastically elegant machines we see in nature is just absurd. Take a look at the hand at the end of your arm. Do you really think that happened by *accident*?"

Bryan was good and ready for this. "Hang on, bro, let's think about this for a minute. Let's say you've got 500 million falcons living around 20 years each and dying over a span of 100 million years—that's trillions of falcons, right? Don't you think that in that huge span of time, over that vast population, one would inevitably develop a helpful feature by accident every now and then? Features that simply weren't there before? Like a new eye muscle that helps them focus and see their prey better?

"Why couldn't that happen? Wouldn't it almost *have* to happen? Then all you need is natural selection to kick in, and the better ones finish off the inferior ones, and you get better falcons."

That question—*Wouldn't evolution almost* have *to happen?*—would not let go of me, no matter how much I tried to shake it. If this were true, it would completely alter the way I saw everything around me. I would never think about my hand at the end of my arm in the same way.

I didn't see a conflict between my religious views and some notion of evolution. I didn't necessarily think evolution was fundamentally incompatible with faith; after all, the Genesis story read, "And God said, 'Let the land produce living creatures according to their kinds: the livestock, the creatures that move along the ground, and the wild animals, each according to its kind'" (Genesis 1:24, ref. 901). If the land produced creatures, then exactly how they all came about wasn't too specific, and in fact Genesis 1:24 didn't sound all that different from evolution to me. I was open to the evidence, wherever it might lead.

But I'd never looked all that deeply into it. I knew there was much I didn't know. The last thing I wanted to do was get into an argument with someone about evolution, a branch of science I knew little about.

But here we were, arguing about it anyway, and the question he raised pierced far deeper than falcon ancestry: What if life itself could arise purely by accident? What if nature was a "blind watchmaker," steered by nothing but random copying errors and blind, pitiless selection?

What if all the sophistication and beauty of life could be reduced to a simple Darwinian formula?

RANDOM MUTATION + NATURAL SELECTION + TIME = EVOLUTION

If creatures inevitably accumulate accidental changes in their DNA, if bad changes get weeded out by "survival of the fittest," if some accidents are good and the best prevail, then all you need is time. Enough time gives you the *grand illusion* of engineering and design (111).

Wow, if that were really the case... that would be flat-out *elegant*. And revolutionary! Could natural selection be a nonstop continuous-improvement machine, one that only needs random accidental changes as inputs, and delivers endless diversity and upward progress as the output? Confirmation that such a thing was true would be a revelation for me. To think that blind forces could engineer greater machines than we can even imagine... with nothing but chance at the controls!

This idea shook my entire conception of the world, and my belief in a universe imbued with meaning and purpose. The possibility that Bryan might be right rattled me at the deepest possible level. What if my intuitions about the world were entirely false? What if everything I believed was wrong?

The rest of our time together in China was spent bickering back and forth about divisive issues. As I boarded the plane home, I wondered where this journey would take Bryan.

Even more terrifying, I wondered where it would take *me*.

I realized through these conversations with my brother that at a primal, intuitive level, my belief in God was connected to my sense that the cosmos and the human body itself could not possibly be here without a Designer.

I had lots of time to study the hand at the end of my arm during the endless flight from Beijing to Chicago. I thought about the muscles in my forearm and the ingenious system of tendons that smoothly, silently operates my fingers and joints.

I thought about the nervous system and its extreme fine-tuning. I thought about how, as you stand, you subtly shift your weight so that circulation is never cut off from any part of your skin for more than a few minutes. We all do such things, mostly unconsciously, as glorious systems maintain the human body for 70-plus years. To me, this was design of the highest order.

If a designer was not needed to produce ingenious designs, then there must have been scientific principles of design they never bothered to mention in engineering school. If blind, unintelligent processes could make arms and hands and ears, then was there some principle the biologists knew that we engineers didn't?

I wanted the truth, even if it might destroy my life as I currently knew it. At the core of my being, I knew I could not live apart from integrity; I could not somehow make myself believe something that was demonstrably untrue. I was about to embark on a journey more terrifying and challenging than almost any I had faced.

As soon as my plane touched down, I started hunting for an answer—hard. Scouring the internet, adding all kinds of books to my shopping cart, I delved into the issue with fervor. I had to know. I bought more books. I listened to radio programs, watched videos, lurked in online discussion boards.

My drive came from the belief that there had to be some sort of mathematical formula or underlying foundation that would demonstrate the possibilities of evolution, and show its limitations.

Could highly structured designs really emerge from blind chaos? Was natural selection all you needed? If so, my purpose was to locate the system or process or set of principles that proved it.

Though Bryan and I had been wrestling with all kinds of questions for two years, this latest conversation had pressed me to the very edge. I was teetering.

Every time I took a position, he countered with something persuasive and carefully considered (that's Bryan for you). Little by little my dogmatic certainty about spiritual matters receded and I discovered what remained.

When all the rest was stripped away, the remaining force that kept me from sliding into atheism was my engineering instincts. I'd been an engineer long enough to *know that I know that I know* certain things.

I had more confidence in those engineering instincts than anything else. I thought of Solomon, who said, "In a lawsuit the first to speak seems right, until someone comes forward and cross-examines" (Proverbs 18:17, ref. 901). So I made a daring, perilous, frightening decision.

I was going to let science and engineering answer this question for me.

I promised that if science really told me that no God, no plan, no intentionality was needed for me to have a wonderfully engineered hand at the end of my arm, then I would make a massive, wholesale change in my belief system.

I thought about how my family life could change.

My wife might wake up one day to find an atheist sleeping in her bed. I could end up staying home while she took my kids to church. Would I have to bite my lip, or would I try to enlighten all my friends and family that their beliefs are based on fantasy? Would Thanksgiving dinner turn into a brawl over science and religion with my devoutly faithful relatives? I might lose lifelong friends over this.

That is, IF. IF evolution was really just a matter of time and chance. IF engineering was possible with no engineer.

Was I committing some kind of sin by trusting science more than the Bible?

Maybe so. But even Saint Paul, in all his logic and theology, insisted that God's power is clearly seen in nature: "For since the creation of the world God's invisible qualities—his eternal power and divine nature—have been clearly seen, being understood from what has been made, so that people are without excuse" (Romans 1:20, ref. 901). Paul insisted the evidence is so obvious that there is no excuse for being an atheist. I reasoned that if you can get engineering without engineers, Saint Paul's own words undermine faith.

I decided: *I am going to get to the bottom of this. Even if it costs me* everything. *I just want to know what's true.*

I made a terrifying leap into the void.

What Would It Take to Make Me an Atheist?

I chose to put my religious biases on the chopping block. I imagined seeing, hearing, feeling, and tasting the world from an atheist's point of view. I asked myself, "In order to become an atheist, what would have to be true?"

- I would need proof that random mutations (which are accidental changes in DNA, as you'll discover in chapter 4*) filtered by natural selection really could generate all manner of elaborate structures like eyes and ears. I understood that natural selection has no creative power in itself; it can only act upon something that mutations have already produced.
- I would need to discover a *principle*—a law in math, science, or engineering—that said, "X percent of the time, random mutations are neutral. Y percent of the time, they're harmful. Z percent of the time, they do turn out to be helpful." This would have to have been rigorously tested.
- I would need *proof*, with bona fide laboratory experiments, that all things really needed to evolve was time and chance.

And one last thing:

- Someone would have to show that the first cells and life itself could emerge from the early ocean without any kind of action by an intelligent agent.

Those were the things I started looking for. The atheist worldview, if correct, came with an engineering bonus: You don't need smarts to design great things, you only need lots of time, lots of accidents, and some occasional good fortune.

Are You Ready to Put *Your* Beliefs on the Chopping Block?

I had a deep interest in the truth, as much as it may be grasped by frail human beings. Ultimately, I decided that if God were real, if there were design in the universe, I shouldn't need a holy book or blind faith to know. Design in nature ought to be detectable. Not just through common sense, but also through normal scientific reasoning and observation. If

* Mutations are changes in DNA. The Neo-Darwinian theory (which is explained in chapter 2) assumes that these mutations are random and not goal directed in any way. However, this assumption is narrower than the basic definition of *mutation*, which is simply a change in DNA. In this book, I question whether evolutionary mutations are actually random. In chapters 11–16, you'll see that useful mutations follow precise protocols.

design could not be detected in nature, that would be a strong vote against faith.

I still believe that today.* Yes, I am a Christian, but I have sampled the major viewpoints, tempered my original position, challenged every assumption, and arrived at a new understanding.

I *love* science and technology. I'm weary of dogma and unfounded assertions that become religion-like, for no good reason. I firmly believe that even if you set theology aside, nature speaks for herself.

Is evolution a process of chance and blind chaos, as some famous scientists insist? Or are living things intentional? Do we need to redefine "evolution" to mean *purposeful* and *adaptive*?

These questions are not merely academic; they matter to our very civilization. If evolution requires ingenuity at the cellular level, and not merely chance and selection, this has sweeping implications for medicine, health care, and technology.

If design and intentionality are essential to life, that changes *everything*. It signals a moral responsibility toward the Earth and toward each other. The only thing we'll accomplish by denying it is to dehumanize ourselves and destroy our planet. And if science points to something beyond ourselves, then we can know for sure that we're not just so many billiard balls banging around in the universe. It means man's search for meaning is not just blind groping, but a quest for something that is real.

I've written this book for people like my brother, who are good hearted, love intellectual curiosity, and welcome all information... even if it bruises their belief system. People like that *demand* to have their belief system challenged, because it's embedded in their most treasured values! Unchallenged, unexamined life is not worth living, as Socrates said, and indeed results in a life that has not been lived.

---

* This section makes many Christians very uncomfortable. I am not suggesting that one cannot pursue faith without first having satisfied all intellectual questions. But I am suggesting that whatever you put your faith in should not contradict obvious verifiable facts. In *A History of Christianity*, Paul Johnson wrote, "For Christianity, by identifying truth with faith, must teach—and, properly understood, does teach—that any interference with the truth is immoral. A Christian with faith has nothing to fear from the facts; a Christian historian who draws the line limiting the field of enquiry at any point whatsoever, is admitting the limits of his faith. And of course he is also destroying the nature of his religion, which is a progressive revelation of truth. So the Christian, according to my understanding, should not be inhibited in the smallest degree from following the line of truth; indeed he is positively bound to follow it. He should be, in fact, freer than the non-Christian, who is precommitted by his own rejection" (929, introduction). More about this in appendix 2.

Whether you are an atheist,* agnostic, or traditional Creationist, in this book you'll encounter key scientific discoveries no one bothered to tell you about.

In the pages to come, I put the above questions on the anvil and start swinging the hammer. So if you, dear reader, think that an examined life and an examined belief system is a good thing... then I start shaking things up in the very next chapter.

Why Should I Read an Evolution Book Written by an Electrical Engineer?

Great question. When I began this journey I wondered that myself. After all, I'd spent five years getting one degree. I certainly knew how complex and subtle one field of science could be. I thought, *Do I have to go get another degree, this time in biology, just to understand this stuff*?

I quickly discovered the field of bioinformatics at the intersection of computer science and biology. Bioinformatics explores the deep parallels between genetic information and human-made systems. Nearly every concept I'd written about in my book *Industrial Ethernet* applied in some way to DNA. Evolution could be studied as a software engineering problem!

I read books ranging from evolutionists Richard Dawkins and Daniel Dennett, to the Intelligent Design advocates, to many less popular, highly technical titles. (I summarize the best ones in appendix 3.) I absorbed everything I could about genetics, communication theory, and bioinformatics. As I waded into public and private debates online, I was able to confirm that my facts were in place and, yes, I was perfectly competent to discuss the areas where biology and electrical engineering overlap.

Despite the fact that I wouldn't pass a test on mollusks or retroviruses, my electrical engineering orientation helped me to highlight aspects of genetics with crystal clarity that biochemists tend to ignore. One reason is that, in electrical engineering, theory matches reality better than it does in most other engineering disciplines. For instance, in metallurgy, when you

* Atheists come up a lot in this book. Most of the time I'm referring to ardent, "evangelistic" atheists with Darwin-fish bumper stickers on their SUVs. However, in private, vulnerable conversations—rarely in public—I've encountered another kind of atheist. This is the undeniably spiritual person who has been searching for God his or her entire life and never found him. I had drinks with a gentleman the other day who said "I'm an agnostic or an atheist, because I've been saying 'God, please show yourself' all my life and gotten nothing but deafening silence."

predict the failure load of a steel beam, you're lucky if your guess is within 10 percent. Civil engineers overdesign bridges by 50 to 100 percent just to be safe.

But a model of an electrical circuit or computer chip is often accurate to within 1 percent and sometimes 0.01 percent. Electrical engineering is highly mathematical. Theory must match reality; everything you design has to ***work***. This extreme precision is no luxury. In the real world, lives and livelihoods depend on accurate models. Electrical engineers have high expectations, and as an engineer I was impressed with nature's designs. I was alarmed that people who, so far as I could tell, had never designed and manufactured a single product in their life felt qualified to announce that "design in biology" is absurd.*

As I probed deeper, I grew worried by the lack of rigor in popular evolution books. "Evolution by randomness" was so entrenched as to be rarely questioned, except by heretics. I also found an immense chasm between the version of evolution you find in the bookstore and what practicing biologists understand evolution to be.

Industries become incestuous as they age. They resist change because change threatens the status quo. Since ***all*** professions are run by good ol' boys clubs, innovations almost ***never*** come from the inside.

For example, Bill Gates was a complete outsider to the computer business. Larry and Sergey, founders of Google, were newcomers to the search engine game. (Early on, they tried to sell their search technology to Yahoo! for $1 million, but Yahoo! turned them down.)

Fred Smith, founder of Federal Express, was a virgin in the shipping industry. Ray Kroc of McDonald's wasn't a restaurant veteran; he was a milkshake machine salesman. Lou Gerstner, who engineered a turnaround at IBM, had come from Nabisco and American Express. Before Jack Welch transformed GE, he was a chemical engineer.

Mathematician and quantum physicist Barbara Shipman, a University of Rochester researcher, noticed that the shape of the honeybee's dance closely mimics something in physics called the ***flag manifold***. Bee experts had never noticed this before a physicist came along (213). Benjamin Franklin,

---

* The "Salem Hypothesis" is an anecdotal observation, originally by Bruce Salem, that electrical engineers, mechanical engineers, and computer scientists are more skeptical of Darwinism than the average person. John Wilkins (256) cites research showing doctors and dentists also match this pattern. The hypothesis states, "An education in the Engineering disciplines forms a predisposition to Creation/ID viewpoints." My experience confirms this is true. This is an important observation and we will return to this question in chapter 27.

a printer and statesman, discovered that lightning comes from electricity. George Simon Ohm, who discovered "Ohm's law" of electricity, was a schoolteacher. Charles Darwin was a medical school dropout studying for the ministry when he took his famous trip to the Galápagos Islands on the ***HMS Beagle.***

Novel approaches usually come from outsiders. All these people had an outsider's point of view that enabled them to see something to which insiders were blind.

Likewise, I didn't invent or discover anything new. Everything you're about to discover in this book can be confirmed by checking the references. I found the most astonishing story I'd ever heard and nobody was telling the public about it. My outsider's perspective as an engineer enabled me to form connections that a biologist would not necessarily make.

ONLINE SUPPLEMENT

Biomimetics—How to steal nature's secrets and apply them in human technology

www.cosmicfingerprints.com/supplement

(Biologists are also privy to things engineers know nothing about. Engineers have ***much*** to gain by exploring biology. The point is, disciplines need to talk to each other.)

Last, my experience as a search engine marketing consultant was a window into the deeply "Darwinian" world of Google advertising. For over 10 years I've been caught up in the everyday dramas of real businesses evolving against cutthroat competition. Google plays the role of natural selection. Its job is to sort the winners from the losers. Parallels between Google and evolution are considerable.

As you read, pay attention to how many different fields are making contributions to biology.

Notice also how much modern biology is contributing to other fields; the world is discovering wonders in biology that you can personally adapt to technology, business, and real life.

The vast majority of my sources are secular, peer-reviewed research from credible mainstream scientists. Check the references. Verify for yourself that the facts are correct. Enjoy this outsider's critique of science's most hotly debated theory.

CHAPTER 2

Evolution: Truth or Fiction?

We come from the water,
Livin' in the water
Go back to the water,
Turn the world around

—HARRY BELAFONTE

A WHALE-WATCHING TRIP in New England surprised me with persuasive evidence for evolution.

A few months after my evolution debate with Bryan, my nephew Ben was getting married in Boston. We arrived the day before the wedding with time for a whale-watching trip in Boston Harbor. Sure enough, we spotted some whales. It was my first time doing such a thing, and it was a joy to see the sun sparkle gloriously on the water as our boat bounced through the early-September waves.

I visited the Whale Center of New England next door and discovered one of the most compelling cases for evolution I've seen. The Center had mounted a large whale skeleton for visitors to examine. The exhibit mentioned something I'd read about: "vestigial whale feet."

Near the back of their bodies, whales have a very small set of bones, not even attached to the skeleton, which are obviously a set of minimally formed feet and legs. I'd never been quite sure what to think about whale feet. Now I was confronted with this real-life skeleton.

Dorudon atrox ("Spear-Tooth") skeleton in the Sant Ocean Hall of the Smithsonian National Museum of Natural History in Washington, DC. Near the back you can clearly see "whale feet" bones. In a live animal, these bones are suspended in its hindquarters, unconnected to the remainder of the skeleton.

Sure enough, this whale had a tiny set of legs, folded up near the back of its body, disconnected from the rest of its skeleton and suspended in the flesh—remnants of an earlier ancestor having been some other type of mammal.

If animals have body parts and extra hardware they don't need, where did that leave me and my engineer's intuitions about living things?

A human engineer, if designing a whale from scratch, would leave out the feet entirely rather than include very small feet tucked inside the whale's body.

At this point in my journey, I preferred the Intelligent Design view of the world. But the exhibit regarded the whale legs as evidence of evolution. The evolution beast had its claws in my skin and wouldn't let go.

As I drove back to my hotel, I pondered whale feet from an engineer's point of view. I asked myself, "What would it have to take for the whale's DNA to still hang on to those legs, but make them so much smaller?"

There was something very, very interesting about this. The whale's entire leg assembly was shrunken down, apparently to scale. The legs on each side were symmetrical and it looked like all the parts necessary to function were still there. Just smaller.

It was like someone twisted the "Size of Legs" dial from "10" down to "1"!

That in itself is a very tricky engineering problem. How do you hang on to most of the instructions, but change *some* of the directions to make the legs smaller?

If this had been a computer program, rewriting that program to build smaller legs would be no trivial task. I thought of the welders and robotic assembly lines at the Ford plant on the South Side of Chicago. What would it take for those robots to start making a tiny little spare tire, instead of a regular-sized one, and install it in the trunk?

Everything would have to scale down in proportion—the rims, tires, nuts, and bolts—which meant a lot of precision programming. Now that I thought about it, shrinking the legs and retaining them was remarkable all by itself.

The genome didn't just delete lines of code and make the legs disappear entirely. It looked as though the adaptive program was trying to hang on to valuable inventory. It seemed almost... conservative. As though it knew it might need those legs some time in the future, and so resisted deleting them. Who knows? Maybe those bones still serve some unseen function now.

An evolutionary algorithm that could hang on to those legs, compressing them to occupy minimal space and resources... that seemed anything but purposeless and accidental. Did an adaptive program possess a kind of intelligence of its own?

Now I was *really* intrigued. I decided to investigate the idea of seemingly unnecessary parts further.

Mole Rat Eye Covered by Skin

The blind mole rat's eye is completely covered with skin. In theory, the eye could be functional, but since it's covered by skin, it can only sense light and dark (229). It doesn't make sense that this species would be uniquely designed this way. It would not seem that a designer with infinite resources would put a fully functional eye underneath a flap of skin instead of fabricating a simpler eye.

Therefore, it seems as though the mole rat is an assemblage of off-the-shelf parts, not a unique and special creation. To most people, an evolutionary explanation would seem more sensible.

If whales and blind mole rats are descended from other mammals, then it might seem to follow that humans are merely primates. Lots of people find this offensive. It's embarrassing enough to admit dad was an alcoholic, let alone announce that great-great-grandpa was a baboon. It's one of the many reasons why religious people don't want to believe in evolution.

Has anyone *not* heard some preacher or pundit rail against this?

The blind mole rat has a functional eye, which is entirely covered with skin. It's not uncommon to find underutilized parts in human designs, too; when you do, it's because the part was readily available and it was too much trouble to design a simpler one.

LifeSite News reported an effort by London Zoo to reinforce the evolutionary view:

> In this very embarrassing zoological charade three brawny men and five shapely women have volunteered to strut around for a few days in their underwear and strategically placed fig leaves in an enclosure that was once the home of one or another species of bears. In the immortal words of the Zoo spokeswoman, the purpose of the exercise is that "*Seeing people in a different environment, among other animals... teaches members of the public that the human is just another primate.*" (127, emphasis mine)

Not to be outdone by those who object to ape ancestry are those who point out our capacity for cruelty. Eric Hoffer, the American philosopher and writer, said, "The pre-human creature from which man evolved was unlike any other living thing in its malicious viciousness toward its own kind. Humanization was not a leap forward but a groping toward survival" (124).

Should we be offended at the idea of evolution? The answer for me came from a very strange place: sex advice for insects.

Sex Advice from a Praying Mantis

Bryan brought home a hilarious book called *Dr. Tatiana's Sex Advice to All Creation*, a biology book written like a newspaper advice column. One chapter starts like this:

> Dear Dr. Tatiana,
> I'm a European praying mantis, and I've noticed I enjoy sex more if I bite my lovers' heads off first. It's because when I decapitate them they go into the most thrilling spasms. Somehow they seem less inhibited, more urgent—it's fabulous. Do you find this too?
> —I Like 'Em Headless in Lisbon

Dr. Tatiana replies:

> Some of my best friends are man-eaters, but between you and me, cannibalism isn't my bag. I can see why you like it, though. Males

> of your species are boring lovers. Beheading them works wonders: Whereas a headless chicken rushes wildly about, a headless mantis thrashes in a sexual frenzy…

In her own funny way, Dr. Tatiana makes it obvious why so many people are so squeamish about evolution. If all we are is animals, we can merrily rationalize every conceivable form of aberrant, grotesque, inhumane behavior.

But think about this for a minute. Can we *ever* justify our own behavior by claiming that animals do it, too?

Dr. Tatiana's book makes it plain: Of course not! You could use her book to rationalize…almost anything!

Ancestry and morality are different questions. The first deals with the survival of the fittest and the second with the survival of all.

There's an episode of *The Big Bang Theory** where Leonard is having a moral crisis:

LEONARD: I am having a moral crisis.

SHELDON: Well, if it is of any help, I have read all the great moral philosophers, including Dr. Seuss.

LEONARD: Oh, what the hell. I am supposed to go out with that girl from the comic book store, Alice, but I do not know if I should because I am going out with Priya. But she is in India.

SHELDON: All right, so the topic at hand is sexual fidelity. Probably we won't be relying on Seuss here. Although, *One Fish, Two Fish, Red Fish, Blue Fish* might be surprisingly applicable. Go on.

LEONARD: Well, they say at the end of your life you will regret the things you didn't do more than the things that you did, and I am pretty sure that Alice is the stuff I want to do.

SHELDON: You know, the German philosopher Friedrich Nietzsche believed that morality is just a fiction used by the herd of inferior human beings to hold back the few superior men.

LEONARD: That actually does help.

SHELDON: It is worth noting that he died of syphilis.

We humans are forever trying to escape the clutches of our own animal instincts. It's a defining quality of what makes us human and

* *The Big Bang Theory*, "The Good Guy Fluctuation," 5-7.

different from all other species. All laws, morals, and ethical standards exist to protect the weak from the strong, to save us from savagery, so that more than just the strongest survive. A society that is *not* Darwinian is the very definition of civilization.

Whenever we talk about living in an evolved society—one that rejects slavery; cares for the sick, the old, and the poor; a world where we help handicapped people get jobs; a world that is "kinder and gentler"—the word *evolved* means the exact *opposite* of what it means in Darwinism.

If you're looking for a place to ground human rights and dignity, modern ideas about equality and social justice, you won't find it in biology. If you wish to rid the world of racism, Mr. Darwin will be of no help. Those are moral questions. That's why the moral struggles of humanity and the engineering problem of evolution are separate domains.

Sure, animals live in groups and share resources with each other much like we do. But there are clearly many situations where altruism has high cost with little or no direct benefit to everyone else. Biology cannot tell you why taxpayers should support a sick stranger a thousand miles away, or why the opportunity to vote is a sacred human right.

Because morality doesn't come from biology or chemistry—or the sexual fantasies of praying mantises—I couldn't see any good reason to be offended by common ancestry. In fact I found the idea tantalizing. (*Hint*: Spirituality is the thing that distinguishes us from animals. It doesn't come from our bodies, it comes from our spirits. See appendix 2 for a much more complete treatment of this sensitive and important topic.)

Once we're able to separate the two, morality and science, we're able to look objectively at the evidence.

Evidence for Evolution: Genes Shared by Humans and Primates

There are small bits of data (pseudogenes) that are shared only by humans and primates, and found nowhere else in the animal kingdom (126). Let's take a sentence in the U.S. Declaration of Independence as an analogy. Out of all the copies of the Declaration, suppose you found two separate examples that both read, "We hold these truths to be self-evident, that all men are created equal, that they are endowed by their Creator with certain unalienable Rights, that among these are Life, *Literacy* and the Pursuit of Happiness."

You might interpret the change as a spelling error, or you could ask why the person copying it chose "Literacy" instead of "Liberty." (Maybe the editor was a librarian.) Either way, for *both* copies to leave out the "-bert-" and insert "-terac-" instead is highly conspicuous. You would logically conclude that both versions were copied from the same document.

The odds are virtually zero that different sources of these passages would include the same alteration, when everything else about the copies was exactly like the original. You would not assume they were two separate copying errors that happened to be identical.

The pseudogenes that humans and primates exclusively share are identical mobile DNA elements (see chapter 11 on Transposition). This is like finding multiple, unique, identical passages in two ancient documents (602). A historian would naturally conclude common ancestry.

This does not *prove* humans and primates had a common ancestor; they also could have had a common designer. If you're wary of making comparisons between humans and primates, I understand. But as I explained, common ancestry didn't offend me. As an engineer, I found myself intrigued with the common ancestry view *because it presented an utterly fascinating engineering problem.*

If God made whales and other mammals with identical parts, that was one thing. But if living things had the innate ability to change their own genomes and generate new species, if the whale's ancestors had the capacity to transform into another species, that was fantastically more impressive. If evolution were true, *God could teach us more principles of engineering through nature* than if it weren't. If evolution makes you uncomfortable, trust me, I understand. But stick with me, because this will make more sense as the story unfolds. Yes, I am suggesting that you can understand God better by studying evolution!

Returning to the whales: As the whale continues to adapt, could it use those feet yet again? Is the whale saving its feet for a rainy day?

The whale's feet, the blind mole rat's eyes, and the shared DNA sequences between species were powerful evidence for some form of evolution.

As I continued to research this, it became clear to me that evolution most definitely appeared to have happened.

But how?

I started hunting through a pile of research from the 1950s to today. I resolved to find some kind of mechanism that would explain it all.

CHAPTER 3

Confessions of a Science Geek

In nature's infinite book of secrecy
A little I can read.

—WILLIAM SHAKESPEARE

WHEN I WAS 13 YEARS OLD, I started building stereo equipment, mostly speakers. By the time I was 17, I was running a small company from my parents' garage and selling my designs on the showroom floor of a local audio dealer. As a senior in high school I was producing speakers that competed favorably with major brands like KEF, Boston Acoustics, and Bose®.

But as a high school student, I was acutely aware of my limitations as a speaker designer. I always had to tinker to get the sound I was looking for. I couldn't predict what my designs would do in advance. Sometimes my hacks would work but I didn't know why. That's because I just didn't have the math background needed. I had to rely on charts, handy-dandy formulas, rules of thumb, and trial and error. Real engineers possessed much more powerful tools. They could predict what would happen with surprising accuracy because they understood all the underlying principles.

I'll never forget an equipment review in *Audio* magazine in my senior year of high school about speaker designer John Dunlavy. Dunlavy was an accomplished communications engineer. He designed his speakers by modeling them as a radio antenna array. Hardly anyone else in the

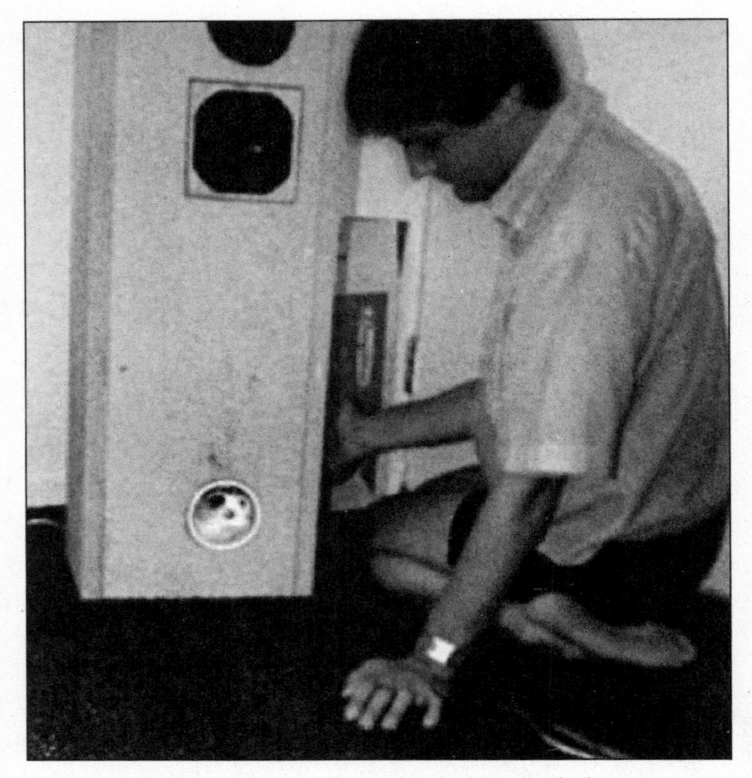

Playing peek-a-boo with my cat Timi, who's looking out the bass port in an unfinished new pair of speakers.

speaker business was doing anything that sophisticated. Most companies were stuck in the usual and customary, but Dunlavy was bringing fresh knowledge from the outside.

His flagship speakers, the Duntech Sovereigns, sold for $15,000 per pair at the time. The *Audio* reviewer described them as "sublime." They beautifully reproduced waveforms that 99.9 percent of all other speakers mangle. To this day, they're considered one of the greatest speakers ever made. Vintage units still fetch $7,000 on the used market. The magazine also described Dunlavy's breakthrough design philosophy.

I was totally inspired. I said to myself, "I want to do *that*." So I enrolled in college to study electrical engineering.

By the time I was a junior, I'd made strides toward that goal. I took extra coursework in communications, control systems, and electromagnetic waves, because those disciplines overlapped with audio and acoustics. Finally, with 90 credit hours of math, physics, and engineering under my belt, I knew how to begin with elementary physical properties like the density of air and current in a wire, and how to predict the exact sound waves a complex acoustical system would produce.

It's hard to explain and it's obviously very geeky, but there is something deeply satisfying about solving a problem at its very roots. At last, I truly understood acoustics from the ground up. This gave me the ability to deal with complex, messy problems with a new level of ease.

One of my term papers was an analysis of a type of speaker called a "Transmission Line." It's very similar to the famous Bose Wave Radio. I derived the math and modeled its behavior on a computer. My professor gave me an "A" on my paper. I was now doing the exact same thing John Dunlavy had been doing five years before: adapting models used for radio waves to study sound. I was running with the big dogs now. Exhilarating.

That's how I discovered early on what it feels like when you're flying by the seat of your pants with no solid foundation. It's the same whether you're working in engineering, acoustics, or evolution. I also found out what it feels like when you finally figure out what you're doing.

Bridging Electrical Engineering and Biology

Watching the Darwin–Design ping-pong ball go back and forth with no clear sense of how to judge between the two sides reminded me of building speakers before my engineering degree. Sure, I could go with a snap judgment. But experience had taught me that snap judgment often gives you the wrong answer. The right fundamental engineering principle, however, gets you the right answer, sometimes easily.

Flashing forward from my science-geek roots to today, I knew if I understood evolution at that same level, I could finally make sense of it. I could bridge two disciplines that usually stay separate, namely electrical engineering and biology.

Evolution, the Universal Acid?

In my argument with Bryan, I was shaken by the profound implications of the power of blind, unguided evolution—*if it worked the way the Neo-Darwinists said it did.* There was no design; there was only the *appearance* of design. There was no purpose; there was only the *illusion* of purpose.

Such an idea, if true, would burn through everything I had ever thought I believed.

Tufts University philosophy professor Daniel Dennett encapsulated this belief in his book *Darwin's Dangerous Idea.* Journalists described Dennett as one of the "four horsemen of the apocalypse"; along with Richard Dawkins, Christopher Hitchens, and Sam Harris, he was among the world's most influential atheists. These brave adventurers were rapidly dismantling thousands of years of religious dogma. Checkmate was just around the corner, evolution their chess queen.

Dennett put it better than anybody: "The outlines of the theory of evolution by natural selection make clear that evolution occurs whenever the following conditions exist: (1) variation: there is a continuing abundance of different elements (2) heredity or replication: the elements have the capacity to create copies or replicas of themselves (3) differential 'fitness': the number of copies of an element that are created in a given time varies" (116).

He described Darwinism as a "Universal Acid": "There is no denying, at this point, that Darwin's idea is a universal solvent, capable of cutting right to the heart of everything in sight. The question is: What does it leave behind?" (116)

Dennett was issuing an unmistakably clear, transcendent engineering principle. The simplification of all simplifications. *But was it really true?*

Was anyone exploiting this Darwinian shortcut to design toasters or race car engines or computer software? Were engineers at Microsoft generating billions of random mutants, letting natural selection perform the hard work of culling and testing? I wasn't sure. But I was pretty sure I would find out. Just as I had gotten to the bottom of physics in speaker design, I knew I could reduce evolution to a set of core principles. I could reach the bottom of this mystery, too.

PART II

THE NEO-DARWINIST DILEMMA

CHAPTER 4

Pity the Fruit Fly: Testing Randomness

What you do to me
It's just like murder
In the first degree

—DOKKEN

THERE HAD TO BE a way to be certain that Darwinian evolution was true.

In engineering shorthand, the formula for evolution was:

RANDOM MUTATION + NATURAL SELECTION + TIME = EVOLUTION

If I could understand the parameters of that formula and its limits, I was pretty sure I could manipulate it to compute the maximum speed of evolution, and perhaps the probability of a new species emerging from a given population. Given certain information, it would tell us how quickly species can develop in what period of time, and what percentage of the time random DNA copying errors help instead of hurt.

Defining the Terms*

Natural Selection:

1. A natural process that results in the survival and reproductive success of individuals or groups best adjusted to their environment and that leads to the perpetuation of genetic qualities best suited to that particular environment.

Random:

1. Proceeding, made, or occurring without definite aim, reason, or pattern: "the random selection of numbers."
2. *Statistics.* of or characterizing a process of selection in which each item of a set has an equal probability of being chosen.

Mutation:

1. *Biology.*
   a. A sudden departure from the parent type in one or more heritable characteristics, caused by a change in a gene or a chromosome.

In almost all popular literature, the standard explanation for evolution is that it is driven by random copying errors in DNA, which produce mutations of genes. These are said to occasionally change the creature in useful ways. Given enough time, whale feet could be folded up, compressed, and "put in the trunk," so to speak, or giraffes' necks could stretch to reach taller branches. Many scientists insist that's all it takes (105, 114, 136).

You'll easily find a hundred biology books that essentially say what the University of California, Berkeley, "Evolution 101" web page says: "Mutations are Random. The mechanisms of evolution—like natural selection and genetic drift—work with the random variation generated by mutation" (136). They go out of their way to emphasize that there is no plan or purpose behind these changes.

* Each definition accessed online January 13, 2015.
Genetics Home Reference, s.v. "natural selection," http://ghr.nlm.nih.gov/glossary=naturalselection;
Dictionary.com, s.v. "random," http://dictionary.reference.com/browse/random;
Dictionary.com, s.v. "mutation," http://dictionary.reference.com/browse/mutation

These mutations produce variations of plants and animals. Natural selection sorts the winners from the losers and that's how life evolves. The implications are profound, if that's true.

But I was skeptical that random mutations were actually beneficial. It would be tremendously useful if this were true, but it didn't jibe with my experience as an engineer. In fact, accepting "random mutation theory" as true opened up a Pandora's box of questions:

- How often are random copying errors harmful?
- How often are they neutral?
- How often are they beneficial?
- How fast can new species evolve through this mechanism? Can you observe it in real time, or does it happen too gradually?
- How effective is natural selection at refining nature's designs?
- How many "bad designs" does natural selection successfully get rid of? Is what they say true, that 97 percent of our DNA is "junk"? (Many biologists insist most of our genes are leftover evolutionary garbage, which we might logically expect to be true if human beings are literally an accumulation of millions of copying errors that barely managed to "make the cut.")

In science, until you have numbers, processes, and successful experiments, you have nothing. So, to start, I hoped to find experiments where someone had produced random mutations in plants or animals and observed evolution in the lab. Maybe they were even able to accelerate evolution by tailoring the conditions of the experiments. Maybe they had found an "optimum" mutation rate.

It didn't take very long to locate those exact experiments.

Dobzhansky's Fruit Flies

Theodosius Dobzhansky (1900–1975) was a highly regarded geneticist and evolutionary biologist, and a central figure in the field of evolutionary biology for his work in shaping modern evolutionary theory. He famously said, "Nothing in biology makes sense except in the light of evolution."

He deliberately induced mutations in fruit flies (*Drosophila melanogaster*) by exposing them to radiation.* Fruit flies are one of the most thoroughly studied organisms in biology and are perfect for an experiment like this because they breed fast, gestating in two weeks. Dobzhansky could simulate the equivalent of 600 years of evolution in only 30 years.

Several papers from the 1930s (122, 123) report that, at the time, biologists believed radiation might be responsible for genetic mutations, which "are the grist of the natural selection mill with the resulting evolution of new forms" (122).

This hypothesis made total sense. Radiation reliably produces DNA copying errors. That means you can even throttle the rate of mutations by adjusting the radiation level. If evolution is driven by DNA copying errors, and if natural selection kills off the inferior creatures and leaves the improved units, this should work!

Many early scientists embarked on this exact path. An early pioneer, Thomas Hunt Morgan, began radiation mutation experiments in 1910 (132); Dobzhansky labored with similar goals for much of his life. Dobzhansky wasn't the only one. Richard Goldschmidt, who is considered the first scientist to integrate genetics, development, and evolution (121), conducted similar experiments on moths. He was the first scientist to practice evolutionary development, or "evo-devo" (117), which is now a major field of study.

So what did they find?

No New Forms Were Created

After decades of effort, these experiments produced every kind of defect you can imagine, including mutant fruit flies with legs growing out of their heads where antennae belong (118). A few of Dobzhansky's irradiated populations did reproduce faster than the regular ones. But no new organs or adaptive systems, let alone anything resembling a new species, were generated. Zero progress after 30 years.

Gordon R. Taylor, a journalist who pulls together an impressive range of evolution data and tests Darwin's theory, summarizes these meager results:

---

* Scientists also use the chemical ethyl methanesulfonate to induce mutations.

This radiation-mutated fruit fly has legs growing out of its head where its antennae should be. Radiation provoked bizarre alterations, but never ones that conferred significant advantages to the insect.

> It is a striking, but not much mentioned fact that, though geneticists have been breeding fruit flies for sixty years or more in labs all round the world—flies which produce a new generation every eleven days—they have never yet seen the emergence of a new species or even a new enzyme (676).

In experiment after experiment, the trend was: Mutations generated by radiation do damage and never lead to major improvements. Dobzhansky wrote:

> Most mutants which arise in any organism are more or less disadvantageous to their possessors. The classical mutants obtained in Drosophila usually show deterioration, breakdown, or disappearance of some organs.

> Mutants are known which diminish the quantity or destroy the pigment in the eyes, and in the body reduce the wings, eyes, bristles, legs. Many mutants are, in fact lethal to their possessors.
>
> Mutants which equal the normal fly in vigor are a minority, and mutants that would make a major improvement of the normal organization in the normal environments are unknown. (119)

Certainly *other* types of mutations might be responsible for evolution. But the fact that random mutations from radiation did not help evolution seemed very significant.

Hanson and Heys (123) reported that the incidence of lethal mutations was directly proportional to the radiation level. They lamented, "We are still at a loss to account for the majority of [useful] natural mutations, and the question of the major cause of variation in organisms remains unanswered."

Fruit fly radiation experiments were extremely useful in one respect: They revealed which genes control various aspects of development. By checking which coding sequences had been damaged, Hanson and Heys were able to map birth defects to certain parts of the genome.

You may be thinking: Maybe the fruit fly has evolved to its optimum state, and that is why any mutation is harmful. But if that's true, why did other species have the same problem? Goldschmidt failed doing the same thing with moths (120), and all kinds of other plants and animals have been subjected to radiation (309, 639). The result in all of these experiments was always the same: frustrating proliferation of damaged mutants.

Self-Repair and Adaptation

I did find a handful of radiation experiments that showed mild, favorable signs of evolution. One was conducted by Francisco Ayala, a student of Dobzhansky.

He found that if he exposed fruit flies to very modest levels of radiation, eventually he got...radiation-resistant fruit flies. Adaptation kicked in, with self-adjustments made by the fruit flies' own cellular machinery (102). His discovery challenged the traditional belief that evolution has no goals. David Stadler and Richard Moyer discovered that modest radiation applied to fungus caused its gene repair mechanisms to kick in (671).

(As you'll see in future chapters, cells employ very sophisticated DNA repair mechanisms. Cells devote extraordinary machinery to detecting and repairing copying errors and to resisting radiation and other kinds of external damage.)

Amazingly, Stadler found that when he "warned" the organism with an initial dose of low-grade radiation a few hours before his main experiment, the organism would "brace itself" by switching on its repair systems! The noted geneticist Evelyn Witkin got similar results with bacteria in the 1940s: She stimulated mutations with ultraviolet light, and instead of mutant bacteria, she got mutation-resistant bacteria. She was so surprised, she made it her PhD thesis (309).

While some papers reported very slight, questionable improvements such as these (134), *I never found a radiation mutation experiment that definitively produced a useful new feature that wasn't already there.*

I hunted for a Darwinian justification for this. Surely someone had a systematic explanation for why these experiments didn't work the way they were supposed to.

Strangely, I could not find one. Major pro-evolution websites like Talk-Origins and Infidels were suspiciously quiet about it. The closest thing to an explanation I could find was "30 years isn't nearly enough time."

Thirty years is admittedly only an instant in geological time. But much later I would find out, through completely different kinds of experiments, that a few decades is more than enough to produce a new species multiple times. In later chapters, you'll see many fascinating experiments where new mechanisms do develop in the lab—fast.

New organs, adaptive traits, cooperative mergers, and new species most definitely are observed today, sometimes in a single generation. However, radiation-induced mutations always appear to be neutral at best and usually harmful. (In chapter 11 you'll discover an altogether different way of applying radiation that opens up a whole new perspective on these fruit fly experiments.)

Could that mean that when fruit flies adapt, they don't transform through random copying errors, but instead by some other mechanism? My mind burned with curiosity. What other kinds of mutations might exist?

Clearly, these experiments showed glaring omissions in Darwin's theory of evolution. The "random mutation + natural selection + time = evolution of new species" model had left out something very big.

A functional theory of evolution would require some other system to get the kind of mutations they were looking for, since "random" wasn't cutting it. Could there be specific types of mutations that performed certain evolutionary operations? Perhaps fruit flies don't progress by random copying errors, but by some other formula? What would that formula look like?

I needed to find out.

Mendel, Population Genetics, and Gene Flow

Population genetics is a branch of biology that deals with the statistics of genetic variation in large numbers of plants and animals.

When populations become separated by migration patterns, mountain ranges, and weather patterns, the genes that are advantageous in one population may not be helpful in the other. As this happens, the populations diverge. But later those same populations can be reintroduced to each other, and the genes of one population may quickly be accepted by the new, larger group. For example, if you introduce finches with long beaks to a population of finches with short beaks, and if it's advantageous, long beaks may rapidly spread through the entire population. This is called *gene flow*.

This is the outworking of Gregor Mendel's laws of genetics within the constraints of natural selection, the statistics of which were worked out in the early 20th century by Sewall Wright, J. B. S. Haldane, and R. A. Fisher (128). Population genetics is a major pillar of Neo-Darwinism because it explains how, once a trait exists, it can quickly proliferate within a given population and get "locked in."

Population genetics is distinctly different from the random mutations in fruit flies I'm talking about in this chapter. On a day-to-day basis, almost all biologists think in terms of genes and networks of genes and deal with what already exists. Mendel's work applies to traits that already exist, rather than the development of new ones. Once genes and traits exist, the laws of genetics and sexual reproduction go to work.

I was already aware that genes rapidly spread through populations when Bryan and I were arguing about falcons. I knew the laws of genetics explained how genes combine to give you blue eyes or brown. But these genetic rules don't tell you where brand new genes came from in the first place. We're asking the question, "Can *completely random* mutations generate new features and genes in the first place?"

Fruit fly radiation neatly isolates this question. It puts the random mutation hypothesis on the chopping block . . . and demolishes it.

CHAPTER 5

Eureka! Information Theory!

The impulse is pure
Sometimes our circuits get shorted
By external interference

—RUSH

AT THE BEGINNING of this journey, absorbing books and scouring websites, I was gasping for breath in an ocean of confusion and frustration. Biology was such an immensely complex subject, and the debates were so charged with emotion.

But the worst part was having no sense of anything solid. The waves of uncertainty tossed back and forth and I rode them like so much flotsam and jetsam. I didn't have a compass. I didn't even have a boat.

I needed something that would make the whole topic less "squishy." I thought probability and statistics might help, so I started asking questions like, "If a hacker has to try 100 billion combinations, on purpose, to guess one eight-digit password, then how long would it take for random copying errors to produce an eye?"*

I soon found lots of people cry foul as soon as you bring up statistics in evolution debates. Some simply refused to entertain such questions

* An eight-digit password with upper- and lower-case letters and numbers has more than 10^{14} (100 trillion) possible code combinations. The bacterium *Mycoplasma genitalium*, widely studied because it has one of the smallest genomes, has 582,970 base pairs. That makes 10^{23} (100 billion trillion) possible code combinations. The human genome, with 3 billion base pairs, has 10^{38} possible code combinations.

at all. Occasionally someone would remark, "Obviously we're here, so it doesn't matter what the chances are."*

I still needed something that I, as an outsider to biology, could use to finally lay my hands on something solid. It turned out that thing was the patterns in DNA.

Darwinism Versus Neo-Darwinism

The term *Darwinism* formally refers to the theory of evolution as Charles Darwin expressed it in *On the Origin of Species* in 1859. He postulated that small variations in organisms over vast periods of time, filtered by natural selection, were responsible for the development of new species.

Darwin's original theory might be summarized like this:

Gradual Variation + Natural Selection + Time = Evolution

Darwin himself didn't strongly evangelize his belief that the variations, or the mutations, were random. Randomness didn't become dogma until the 20th century. Darwin didn't know where the variations came from. He didn't know about genetics and he didn't know anything about cells or DNA.

At the exact same time Darwin was writing his book, Gregor Mendel (232) did a series of famous experiments with peas where he worked out the basic rules of inheritance—the rules everyone learns in high school biology, how dominant and recessive genes are passed on from mother and father to offspring. Darwin had no knowledge of Mendel's theory. He also thought children could inherit learned characteristics from their parents (108).

* One would surely expect to find a book somewhere called *The Statistical Case for Random Mutations* or something like that, which would methodically demonstrate that an acceptable percentage of beneficial random mutations will inevitably occur. Such a book would be a staple of any evolutionary reading list. After all, one can hardly practice science without math! But alas, I found no such book. The rigorous treatments I did find, such as Fred Hoyle's *Mathematics of Evolution* (125) and a symposium at the Wistar Institute (131) showed evolution via random mutations was exponentially improbable.

Since I'd taken probability theory, I attempted to engage with numerous people on this level. Few had the necessary background to respond properly; mostly I got stonewalling. One guy was completing his master's degree in mathematics at Dartmouth, but he could not make a case for random mutations, either. Since the average person is much more familiar with codes than probability theory, I abandoned the statistics angle in favor of the approach you find in this book. Still, the random mutation hypothesis must obtain statistical validation before it can be legitimately accepted as true.

Neo-Darwinism, by contrast, refers to the *Modern Evolutionary Synthesis*, which was developed in the 1930s and '40s. It rejects the idea of passing learned traits to offspring and adds Mendelian and population genetics, including the nonrandom recombinations of genes through sexual reproduction, which are critical to our understanding of trait inheritance (128). Neo-Darwinism explicitly denies any purpose, prediction, or programming in evolution (129). It is the prevailing paradigm in science today (114).

For simplicity, in this book I will normally refer to Neo-Darwinism and the Modern Synthesis as *Darwinism*. Here, the use of *Darwinism* emphasizes assumptions of randomness. Mayr and Provine's definitive textbook *The Evolutionary Synthesis* begins by saying, "The term 'Darwinism' in the following discussions refers to the theory that selection is the only directional factor in evolution" (128).

In the next few chapters, I ask the question: Does randomness even belong in this formula at all?

Because randomness and absence of purpose are essential to the Neo-Darwinian Modern Synthesis, I do not consider biologists who reject the "random mutation" hypothesis to be Neo-Darwinists.

It would be inaccurate to characterize evolutionary biology as only being concerned with *random* mutations. In fact, when some biologists say "random," they don't actually mean random at all; they mean "not goal seeking" (105). I explore this in appendix 1, "All About Randomness," which many technical readers have said they especially enjoyed.

Making a Living in the Code Wars

I've had the privilege of being right in the middle of not one but three technological revolutions as they unfolded. I got my first email address in 1995. Two years later, as the internet was catching fire, I was working at a hardware/software firm that sold industrial networks. These networks ushered a new age of technology into manufacturing. Suddenly it was possible to get data from anywhere to anywhere else, and everybody was piling on the bandwagon. The competition was furious.

Then there was the astonishing ascent of Google. In 2003 I was invited to a direct marketing industry conference to speak on Google's new advertising platform. Then I published a book, *The Ultimate Guide to Google AdWords*, and it sold well, having been translated into a half-dozen languages.

To me, Google wasn't some mysterious magical machine like some people thought. It was just another communication system, a language device based on 1's and 0's. Networks for manufacturing, then the internet itself, and then the world's most popular search engine: all were just different applications of code.

DNA was no different.

All these technologies had a great deal in common. They were all about language, digital information, and the opportunity to compete in a crowded marketplace. Each had lessons to teach about the other. Living things, businesses, and technologies alike had to upgrade every day, or face extinction. Cell phones, startup companies, and strains of bacteria were evolving in front of our eyes, every day.

Eureka!

In 2002 I'd written a book, *Industrial Ethernet* (708), about networking devices together on the factory floor. I included several pages about the engineering concept "OSI 7-Layer Model." This is a computer science model that represents information in seven stages of encoding and decoding. OSI is central to data organization in computer networks.

Wading through science websites, I came across comparisons between the human genome (DNA) and the OSI 7-Layer Model (301, 500)—and I stopped.

I read the comparison again... and suddenly a thousand connections sparked in my mind. I understood the 1's and 0's of computers; that meant I could understand DNA, too. *Hey, wait a minute,* I thought... *this is all digital code!* This evidence was the framework through which I could verify or deny evolution!

Claude Shannon, an engineer at Bell Labs who had previously developed a new algebra for genetics in his PhD thesis at MIT (247), pioneered an engineering discipline called *information theory*. It offered a framework for understanding both computers and genetics. Seen through Shannon's work, the parallels between DNA and our modern digital world were striking and beautiful. Scientists were finding systems in billion-year-old cells that were the same as technologies we thought *we* invented 30 or 50 years ago! This was exciting because the parallels between computer code and DNA aren't merely analogies. DNA, I realized, is literally code.*

For you to fully appreciate my "Eureka!" moment that DNA is literally code, I need to explain how information speeds along this blue cable known as the Ethernet, which connects your computer to the internet.

Let's go there now . . .

* DNA is literally and not figuratively a code, according to Claude Shannon's universally accepted definition of code in communication systems (320, 321, 326). I cover this in chapter 6. Watson and Crick's landmark discovery of the genetic code in 1953 was that the pattern in DNA is by definition a code (303, 313, 211, 215, 500, 311, 302, 326). DNA, like many human-made codes, also has redundancy (326), error correction (312), checksums (316), linguistic structure (403, 520), and codes layered inside of codes (675). Appendix 4 shows how ASCII (a simple computer text language, used in keyboards, for example) and DNA both encode, transmit, and decode digital information, based on standard definitions in communication theory. Appendix 4 also offers a substantial technology prize for discovery of a naturally occurring code. I explain the nature of codes and DNA more fully in chapters 6–9 and 23.

CHAPTER 6

Russian Dolls: How Information Stacks Up

Call it dumb, call it clever
Ah, but you can get odds forever
That the guy's only doing it for some doll

—FRANK LOESSER, FROM *GUYS AND DOLLS*

THE OTHER NIGHT I watched *2001: A Space Odyssey* with my 16-year-old son Cuyler. Early in the film, Dr. Floyd is comfortably cruising in a large spaceship. He decides to banter with his little girl back home on Earth on a video phone chat. *Dang,* I thought, *it looks just like they're talking on Skype*. Which, to my son, is utterly normal.

It occurred to me that I needed to explain this to him: "Cuyler, this movie was made in 1968, before I was even born. An audience watching this movie would've thought that video phone call was one highfalutin technological marvel." *Star Trek* predicted cell phones. *The Twilight Zone* anticipated amenities like driverless vehicles and radical cosmetic surgery. I was more impressed by this than Cuyler was. To him, it was just normal.

Because we swim in an ocean of information every day, we give little thought to the inner workings of our revolutionary communication tools. But allow me to peel back the layers of what really happens when you do the 21st century's most ordinary thing: send pictures in a Microsoft Word document to a friend on the internet.

As you'll see in a few minutes, how computers exchange information is highly relevant to questions about genetics and evolution.

To send a picture in Microsoft Word, you create a blank document, type some paragraphs, and insert a JPEG photo you took with your cell phone. Then you email it to me as an attachment. I get your email and open the file. What just happened?

The document looks like this:

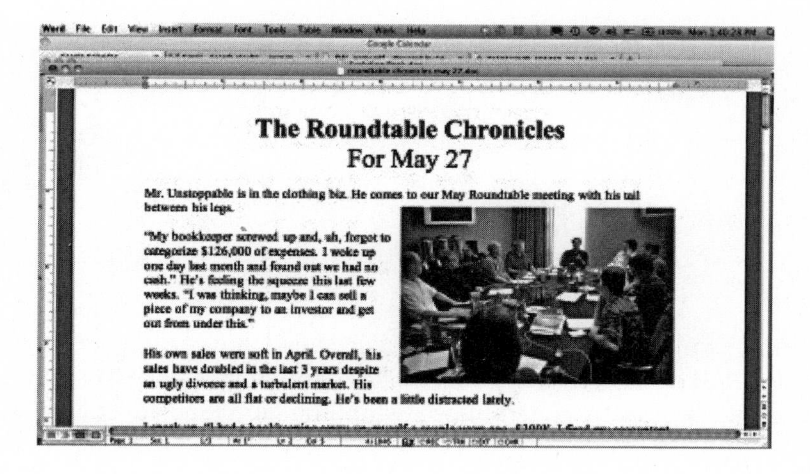

The Roundtable Chronicles
For May 27

Mr. Unstoppable is in the clothing biz. He comes to our May Roundtable meeting with his tail between his legs.

"My bookkeeper screwed up and, uh, forgot to categorize $126,000 of expenses. I woke up one day last month and found out we had no cash." He's feeling the squeeze this last few weeks. "I was thinking, maybe I can sell a piece of my company to an investor and get out from under this."

His own sales were soft in April. Overall, his sales have doubled in the last 3 years despite an ugly divorce and a turbulent market. His competitors are all flat or declining. He's been a little distracted lately.

And when I receive the email, it looks like this, with an icon I can click to open the document:

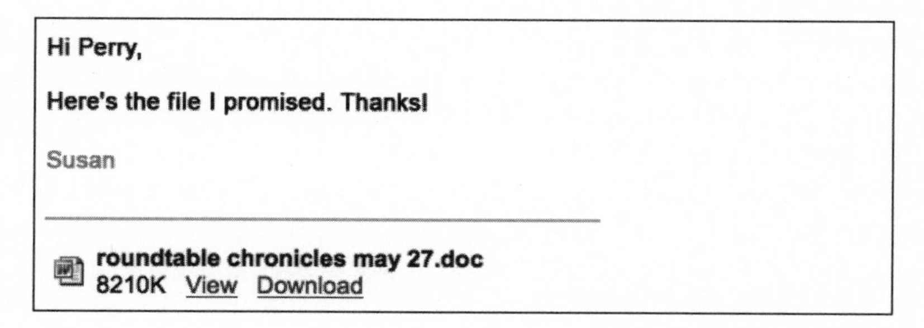

Hi Perry,

Here's the file I promised. Thanks!

Susan

roundtable chronicles may 27.doc
8210K View Download

As simple as this is for us as users, your computer and mine together performed an exquisitely layered encoding and decoding of information.

Have you ever seen those wooden Russian *matryoshka* dolls—where you open the doll and there is another doll inside, and then within that doll, another doll, and so on? The steps of encoding/decoding information work a lot like those Russian dolls.

You open a blank document, write some text, and insert the pictures. Each picture is like the innermost Russian doll—the tiny one in the center.

The "picture" doll fits inside the text, which is the next larger doll. The "text" doll fits inside the Word document doll. The Word document goes inside the email message.

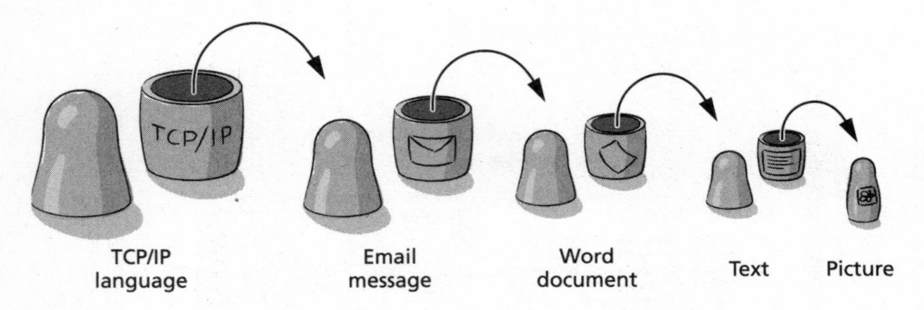

When you press the "send" button, the email goes inside yet another Russian doll called TCP/IP, which is the universal language of the internet. Think of it as the international shipping container for digital data (708).

Let's say you have a wireless network in your house. The TCP/IP message gets rolled inside another Russian doll called wireless Ethernet. And then the digital signals of wireless Ethernet wrap inside another Russian doll, which is a radio signal.

The simple act of sending me a Word doc with pictures in it involves at least seven Russian dolls—seven layers of "message inside a message." Really, there are dozens of layers—I skipped a few just to keep things simple.

When you press "send," your computer stacks all those Russian dolls at lightning speed. Your message speeds through an Ethernet cable out of your house, onto the internet, and comes to me.

When I open your email, the Russian dolls come apart in the exact reverse order. When I open your document, the last Russian doll pops open. Now I can read your document and look at your pictures.

It is important to keep several things in mind:

- A language is a sophisticated code. Every doll represents a different language. Microsoft Word ".doc" format is a special language used by the Word program, email format is a language, and TCP/IP is a language. Your email to me is multiple languages inside of languages.
- As your email comes to me, if any part of the message gets corrupted, it won't just hurt one layer—it will usually destroy *all* the layers (708). It's like chopping your entire stack of Russian dolls in half with an axe. If you're exceptionally lucky, you only splinter the outer two and leave the inner ones intact. But even one tiny data corruption as the packet speeds across the internet will wreck large amounts of information and possibly everything. Ask anyone who's tried to recover a crashed hard drive!
- The dolls have to be unpacked in the reverse order they were packed in. You can never, ever violate this rule! Each doll has to be unpacked by the program that understands its particular language. No other program will work. You edit Word docs in a word processor. You edit emails with an email program. You need to edit pictures in a photo editor. You can't edit pictures in Microsoft Excel. I cannot overstate this point, because in a multilayered system, *any change (mutation) to the code has to obey the rules of that particular layer, while leaving the other layers perfectly intact*. Otherwise, your delicate Russian doll shatters.
- Every layer has mechanisms to check for errors and correct them. When you save a Word document, the Microsoft Word program triple-checks that every single bit has been stored correctly. Software programs, hardware, and networks employ special built-in systems to do this job, called *checksums* and *cyclic redundancy checks* (702). When you save or send your email, the email message with the Word doc inside passes through another set of checks.
- At the end of the day, every single one of those Russian dolls is a single string of 1's and 0's—the alphabet of computer languages. A message inside a message inside a message. The program interprets the message. Without it, the message's meaning can't be understood.

Your computer possesses a vast set of tools for storing, managing, and processing all that data. You can think of each one of those tools as a blade of a Swiss Army Knife.* Each one of those blades has to be applied by the right program at the right time in correct sequence.

The Russian doll analogy refers to the fact that in real systems and real languages, there's a whole stack of encoders and decoders, not just one pair. But digital data is even more delicate than a fine Russian doll, because almost any corruption of any layer will destroy all the layers above it. The extreme fragility of data would prove to be a key to my understanding evolution.

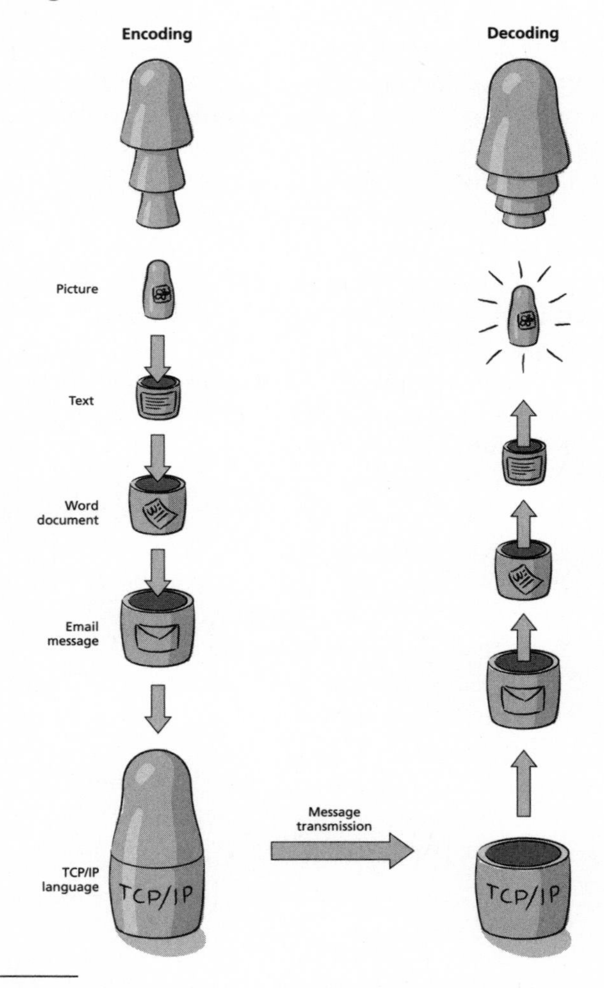

* One definition for "Swiss Army Knife" in the Macmillan dictionary is, "A method or system that deals with situations of all types."

Information Organization in DNA Is the Same as Digital Data!

As I began to study DNA, I saw the striking resemblance to the Russian dolls of emails and internet messages. Data in DNA is also stored in layers: Russian dolls inside of Russian dolls. Here's what I learned:

- DNA nucleotides (ladder rungs in the DNA strand) form codons. *Codons* are groups of three letters, A, C, G, or T.
- Each codon is an instruction to build one amino acid from specific elements. (However, the codons do not *become* amino acids. This is *very* important. The codons are not chemical building blocks; they are *instructions*. A codons gets folded right back into the DNA strand after it is read.)
- Amino acids are chained together according to instructions to form proteins.

Then as we move toward the innermost layers in the Russian doll, they organize larger and larger groups of data:

- Genes specify assemblies of proteins.
- Combinations of genes form an interdependent complex of instructions and a matrix of code elements, packed into chromosomes (675).
- Introns, exons, and transposons are sequences that perform advanced rearrangements of data.

In humans, 46 chromosomes comprise instructions for building the whole body—eyes, ears, bones, fingers, and legs (208).

Just like your computer, DNA packs all of this information into a single string of 1's and 0's; only in the case of DNA, it is a string of A, C, G, and T. Just like digital data, DNA encapsulates that information in layers within layers of instructions.

And just like digital data, the information in DNA is very fragile.

```
x>˘Ó˙Àüà`¨>SíPàÏK\ÜÛıÔÂë"#˘ı>""9¬B?˙ºÁ! ìiÃ      Ça¸°Ò ∞s?¸0p}Ó
(Âq8>ÂÕôÄÄ`E§KòúÁ†§¢T+Õ∑-'•)Ç
®I»HS˘ÛÇ
˝yV<g\óÈr®x-ḯó¸Ÿô¶©Ë%iO˜C!Ãàa»ΩMQÛ¿`ñ%c„¢c«]9>˙∑oœùÈoø˙gS∫¸ß?ST√{ı
´•O®ì=IxlÁ`Äú®J·.ÆÂ?ê`™Tπ                    π|≥8∆'SÈgÈWßKA=éd°zÜ˘∆
€û@¸&äú√RΩIÁPœ„iŸ˘¸:‹}ÕÏÂ\¿‰e˙à(8ä%î-°T5ΩêG"[Y  !¬a1,2œ1gâú≥3È^Õ∫‹
U}T<ß\ó√®yNπªxIG≥¸QQ'PNèewòÍlE0t çE¨„¶CN˝èSÿıIÛ3Ê7/ãC¥?'v1øo◊Íosÿ
3"© ≥#Ja`-êt'†)b«âú3,"Ø˙ıMY]úxK+˜ëQ'ß£'vj¥I! ãMcj}Xı=ïR∂IÙov•©_iÁ
´ﬂw9òÛ>QŒdq˜ûaàZ∏cŸàwÈ¥ßW∂œùÌÌ\o˘˘Î˙€£•V˘ÛÇ
ÈÀX<H\ñÉ∞xPπÃ˘´\+¢%Ï˘o èÏ•≥_¬ÇÒëíöë/‡w˘Sµr]‹r6˙èÊä˘òÍÃûÈ,{WAA"(I√
```


A corrupted file on a computer. This one is intact enough to open, but the program can't read the data inside. It only takes a very few damaged bits to shatter the whole Russian doll. DNA is very fragile! (659)

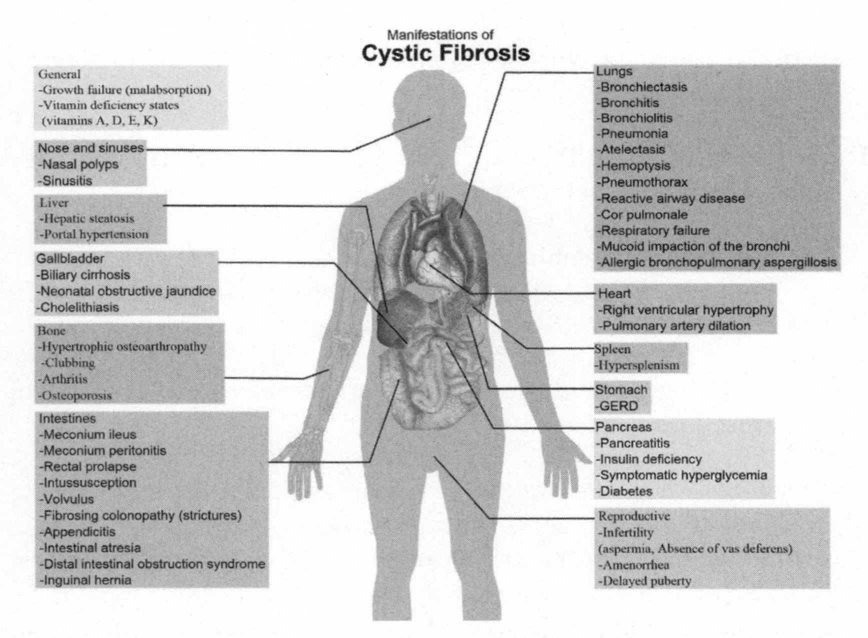

One missing codon wreaks havoc: Cystic fibrosis is caused by a deletion of three base pairs in one gene on chromosome 7. This tiny copying error causes a major birth defect. The effects of cystic fibrosis across 12 different regions of the body demonstrate how many ways a single gene gets used. It also demonstrates the highly interdependent relationships between genes.*

* A software developer might offer an analogy: Hundreds of other genes make function calls to the cystic fibrosis transmembrane conductance regulator gene, and this seemingly minor corruption of a 189-kilobase file compromises the performance of all connected systems.

"Gene" Is a Slippery Word

There is no one precise, universally accepted definition of the word *gene* (643). To most people, *gene* means "one of those things that makes my eyes blue." Indiscriminately calling the many specialized sections of DNA *genes* is kind of like referring to coats, hats, shirts, socks, necklaces, pantyhose, shoes, gloves, rings, zippers, watches, and umbrellas as *garments*. In this book I freely use the term *gene*, but *coding sequence* is much more accurate. Many standard terms in biology are simplistic and misleading.

My fave-rave professor from college, Dr. Robert Knoll, told of touring Turkish castles barefoot because each room's floor had a unique surface with a distinct texture that visitors could experience on the bottom of their feet. He said the Turks had dozens of words for tactile sensations that we have no equivalent for in English. Our vocabulary for the genome is similarly limited. Narrow language limits our thinking.

Russian dolls illustrate the OSI 7-Layer Model, which communication engineers know. I realized that random copying errors could not *possibly* be the driving force of evolutionary change. Why?

Because DNA is also a stack of Russian dolls! You corrupt one tiny bit of data and it grinds the whole doll into splinters; even a tiny corruption in DNA can cause major birth defects (214, 244). You must also obey the rules of each layer's language in order to make useful changes. This is true whether you're talking about "coding DNA" or "noncoding DNA." Coding DNA is direct instructions to build proteins with amino acids. Noncoding DNA contains more complex instructions that regulate development and myriad other processes. Both obey the rules of code (305). (I explore this more fully in chapter 19.)

Just like your computer, cells carry a sophisticated array of tools for reading, writing, editing, and processing DNA. Each tool must be applied at the proper time and at the right *layer*; otherwise information is destroyed and birth defects result.

If I email you a picture of a fender, you can't attach that fender to your car. But I can certainly email you a program that welds fenders on cars. Computer programs build three-dimensional objects, too, not just images on a screen or printer; I worked for years in manufacturing, where complex programs written in "ladder logic" assembled vehicles.

Codes in DNA build three-dimensional organisms just like codes in automotive plants build three-dimensional cars.

Every Message Is Carried by Some Physical Medium

If you send me an email, your message travels to me on a wire. That wire is the "physical layer." If you burn your Word document to a CD instead of sending it in an email, then the CD is the physical, outermost layer instead. The physical layer is the outermost Russian doll. The content and meaning of your document is the innermost doll.

In DNA, the molecule (the double helix) is just the physical layer. Genetic information, like the OSI 7-Layer Model, is layered like an onion.

Driven from the Top Down

Using the lingo of communications engineering, traditional Darwinian theory claims evolution is fueled by accidental damage to the physical layer. By accumulating scratches on copies of copies of copies of CDs, do you think Frank Sinatra's "New York, New York" could transform into U2's "Sunday Bloody Sunday"?

If evolution were true, it would have to be driven from the innermost layer. The conceptual, inner layer is the core of the information. The wire in a computer and the DNA molecule in genetics are the "skin" of the information chain. They are only the surface-level exterior. Because of my communication engineering experience, I know evolution *had* to be organized at a systems level.* In information systems that's how everything works.

Therefore, any evolutionary change would necessarily begin at a systems level within the genome. You can't edit a Microsoft Word document by corrupting individual bits on your hard drive. You always

* Within the OSI 7-Layer Model, "inner dolls" correspond to the top (layer 7 = application layer), and "outer dolls" correspond to the bottom (layer 1 = physical layer). On your computer, you work through the high-level programs (web browser, MS Word) and your actions cascade through the lower layers. We say that evolution is top down, not bottom up, much like we say a company is organized top down, starting with the president and CEO and moving down to vice presidents and managers, etc.

have to start with the innermost Russian doll if you want to change the entire doll. Errors of copying don't do that. They're like trying to make changes to the inner doll by damaging the outer doll.

Those who embrace traditional Darwinism assume that copying errors will occasionally be beneficial, and that when they are, natural selection will do its job and they'll dominate (116). Information technology dismantles that theory, since accidental mutations are not beneficial.

In chapter 9, we'll talk more about copying errors. First let's look at some surprising discoveries about the genetic code.

CHAPTER 7

Why the Genetic Code *Is* a Code, Not Merely *Like* a Code

One humanoid escapee
One android on the run
Seeking freedom beneath a lonely desert sun
Trying to change its program
Trying to change the mode—crack the code
Images conflicting into data overload

—RUSH

CELEBRATING THE 50TH ANNIVERSARY of his co-discovery of DNA, Nobel Prize winner James Watson published a book, *DNA: The Secret of Life*, in 2003. The book opens with his account of unraveling the mystery of this amazing molecule:

> As was normal for a Saturday morning, I got to work at Cambridge University's Cavendish Laboratory earlier than Francis Crick on February 28, 1953. I had good reason for being up early. I knew that we were close—though I had no idea just how close—to figuring out the structure of a then little-known molecule called deoxyribonucleic acid: DNA.
>
> This was not any old molecule: DNA, as Crick and I appreciated, holds the very key to the nature of living things. It stores the

hereditary information that is passed on from one generation to the next, and it orchestrates the incredibly complex world of the cell. Figuring out its 3-D structure—the molecule's architecture—would, we hoped, provide a glimpse of what Crick referred to only half-jokingly as "the secret of life."

We already knew that DNA molecules consist of multiple copies of a single basic unit, the nucleotide, which comes in four forms: adenine (A), thymine (T), guanine (G), and cytosine (C). I had spent the previous afternoon making cardboard cutouts of these various components, and now, undisturbed on a quiet Saturday morning, I could shuffle around the pieces of the 3-D jigsaw puzzle. How did they all fit together?

Soon I realized that a simple pairing scheme worked exquisitely well: A fitted neatly with T, and G with C. Was this it? Did the molecule consist of two chains linked together by A-T and G-C pairs? It was so simple, so elegant, that it almost had to be right. But I had made mistakes in the past, and before I could get too excited, my pairing scheme would have to survive the scrutiny of Crick's critical eye. It was an anxious wait.

But I need not have worried: Crick realized straightaway that my pairing idea implied a double-helix structure with the two molecular chains running in opposite directions. Everything known about DNA and its properties—the facts we had been wrestling with as we tried to solve the problem—made sense in light of those gentle complementary twists.

Most important, the way the molecule was organized immediately suggested solutions to two of biology's oldest mysteries: how hereditary information is stored, and how it is replicated.

Despite this, Crick's brag in the Eagle, the pub where we habitually ate lunch, that we had indeed discovered that "secret of life," struck me as somewhat immodest, especially in England, where understatement is a way of life.

Crick, however, was right. Our discovery put an end to a debate as old as the human species: Does life have some magical, mystical essence, or is it, like any chemical reaction carried out in a science class, the product of normal physical and chemical processes? Is there something divine at the heart of a cell that brings it to life?

The double helix answered that question with a definitive No. (254)

James Watson's statements, if correct, meant life isn't so special after all. If DNA is the product of normal physical and chemical process... *no designer necessary.* For years before Watson and Crick's discovery, some biologists had embraced a mystical theory of "vitalism"—that life possesses some sort of special, immaterial essence that nonliving things lack. Watson was directly countering this belief. Ever since then, many scientists have confidently asserted, as Crick did that day in the pub, that Watson and Crick had solved the mystery of life.

Life, like everything else on Earth, is just chemicals.

Did they really solve it? Or are there more layers to the onion? If you perfectly described all the chemicals and metals you need to make a hard drive, would that explain the software, too? I'll get to that in a minute.

DNA 101

Nearly every cell in your body has a double strand of DNA, a set of instructions for building you. It largely determines the color of your eyes, the color of your hair, how tall you are, your blood type, and literally 100,000 other nuances that make you different from everyone else in the world. All of that information is encoded in just 3 billion base pairs—about 750 megabytes of digital data, or the same amount of data that fits on a compact disc.*

The information in that double strand of DNA is a data packet. It's no different than a string of 1's and 0's flowing into your computer from the internet or stored on your hard drive; it's just stored on a different physical medium.

Computers store digital data as pits on a CD, pulses of voltage on a wire, bursts of light, or magnetic domains on a hard drive. DNA stores digital data in a four-letter alphabet, each letter a nucleotide with a different base: adenine (A), cytosine (C), guanine (G), and thymine (T).

When DNA divides in half along base pairs to make copies for a new cell, complementary base pairs, which are already ready-made by the cell, fall into place and attach to the helix. Now you have two double helix strands of DNA, not one. The cell has made a perfect digital copy.

---

* The current Windows version occupies 16,000 megabytes—20 times more than the human genome. Mac OS needs 5,000 megabytes—that's seven times more space than the human genome. The coding scheme in DNA uses its "hard drive space" very efficiently.

Conceptually, it's no different than duplicating a CD or emailing a Word document to a friend—now she has a copy, too.

Those molecules of adenine, cytosine, guanine, and thymine are arranged in groups of threes, called *codons* or *triplets*. Since there are four different possible letters for each base pair, there are 4 × 4 × 4 = 64 possible triplets, or 64 words in the genetic vocabulary.*

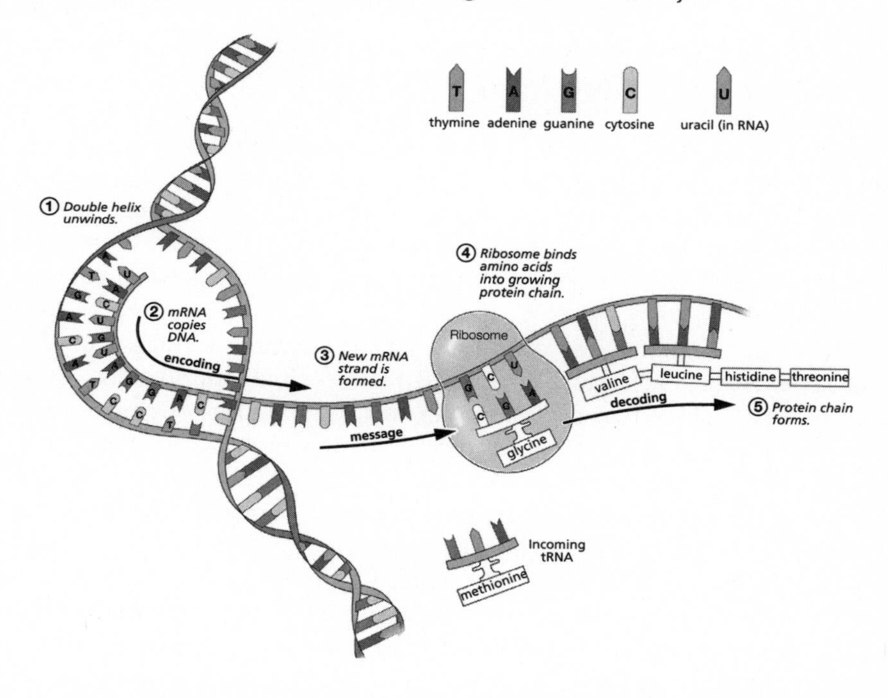

How a cell reads instructions in DNA to build new structures: (1) DNA strand unwinds on the left. (2) A messenger RNA (mRNA) strand is transcribed in the center (message is encoded). (3) The ribosome assembles strings of amino acids on the right by translating (decoding) the instructions in the RNA code. (4) A machine in the cell called a ribosome strings amino acids together to build proteins. After the DNA code is transcribed to mRNA, the helix "zips back up" and the cell coils it back into the nucleus.

These 64 letters form instructions to build 20 different amino acids. That's right—different letter combinations make the same amino acid. For example, GGG makes glycine, but the letters GGC and GGA make glycine, too. Early biologists called this a "degenerate code." The term is

* See ref. 243 or Wikipedia for a table of standard genetic code.

still in use today (205). They didn't really understand this redundancy. Some ridiculed it, believing three different instructions per amino acid to be wasted space.

Months in a Ford plant taught me otherwise, where backup systems prevented downtime, which cost $15,000 per minute. In DNA, these "backup letters" are insurance; it's why the fruit flies were able to resist radiation in some experiments. They insured that a miscopied letter here or there did not result in a birth defect. Without these safeguards, even a smidgen of radiation would've killed the fruit flies on the spot.

In engineering, we call this *redundancy.* When I designed factory equipment, systems had backup systems to make sure that even if one thing went wrong, communication didn't fail. Some of my customers would build a robot with two computers, two communication cards, and two cables, just in case anything failed. It cost *more* than twice as much to install, but the extra expense paid for itself the first time anything broke.

Incredibly, DNA has redundancy built in. Mapping 64 letters to 20 amino acids (instead of just mapping 20 letters to 20 amino acids) creates a 3:1 backup system that guards against single-letter copying errors.

I also discovered this redundancy is extremely optimal. Freeland and Hurst (306) found out this redundancy is so well chosen that if you compared it to a million other random coding schemes, DNA would be the very best one for minimizing errors (306).

Another reason error correction matters so much is that there is not one "genetic code"; there are many. This aspect of the code, the triplet code, is only the most famous one. There are many other codes interlaced on the same strand of DNA. They generate different instructions, depending on where you start and stop reading, and whether you read them forward or backward (660). For example, the differences between a kidney cell and a skin cell are only possible because each cell reads different codes from the same DNA strand. So a single error in DNA will affect many organs in wildly unpredictable ways.

DNA is formally defined as code. Sometimes skeptics wonder if the pattern in DNA is truly a code in the same way that computer programs are code; perhaps it just has some code-like properties. In online discussions, several people have said to me, "DNA only resembles code. Yes, I understand that cells have a sort of 'code' and they 'correct errors' and they 'repair defects.' But in actual reality it's nothing more than chemicals and chemical reactions. Code is only a word that we humans use to describe it."*

Indeed, sometimes these people actually type scare quotes around the word *code*. It's almost like when someone looking back on China's Cultural Revolution might say, "Chairman Mao sent writers and dissidents to work camps to 'reform' them and keep 'dangerous elements' from harming the people."

To shrug off code as metaphorical is to dismiss rigorous definitions and precise scientific terms. DNA's pattern is *literally* code. That's true based on standard engineering definitions (320, 321). This fact is central to the field of bioinformatics, 21st century genetics, and this book.

Why DNA Is a Code

The most common computer code is a language called ASCII. ASCII is a simple scheme your keyboard uses to talk to your PC, in which different combinations of seven bits (0's or 1's) are mapped to the English alphabet.

There are 128 ASCII characters—lowercase and capital letters, numbers, and punctuation marks. "A" is "1000001," "B" is "1000010," and "Z" is "1011010." You can look up the whole table at **www.asciitable.com** or on Wikipedia.

ASCII is a code because when you press the letter "A" on your keyboard, your keyboard encodes it as "1000001," sends it down the cable and into your computer, and your computer then decodes the "1000001" to display the letter "A" on your screen.

In the same way, DNA is a code (304, 318) because the codons on DNA strands are *encoded* into messenger RNA (ribonucleic acid) and *decoded* into amino acids and proteins (326). For example, the base pairs GGG (guanine-guanine-guanine) are instructions that the ribosomes use to make the amino acid glycine.

* One guy said to me, "Codes do not actually exist. They are only abstractions that exist in our own imaginations. In reality, everything is just quarks." If the only relevant property of any object was quarks, then there would be no difference between a Ford Fiesta and Lady Gaga. The view that these things are only mental projections is called *antirealism*, the belief that nothing exists outside the mind. This would mean we have no access to an objective external reality even if it does exist. If you put scare quotes around "code," where do you stop? Does everything in science only occur inside our own heads? Do cells reproduce, or do they only "reproduce"? Are boulders heavier than pebbles or are they only "heavier"?

This process follows the rules of the genetic code. GGG encodes glycine, CGG encodes arginine, AGC encodes serine, and so forth. Note that GGG is not literally glycine, because the GGG nucleotides never end up in the glycine (221). Instead, the nucleotides merge back into the original DNA strand after it is read. GGG are the symbolic instructions to make glycine. The cell's machinery reads these instructions and obeys* them.

This table compares the two codes:

ASCII	Genetic Code
2 bits (1 & 0) 7 bits per letter	4 bits (A, C, G, T) 3 bases per codon
128 symbols (possible combinations)	64 symbols (possible combinations)

A code is the rules of a communication system. The term *code* often also refers to the message itself. To have a communication system, you need four things:

- A *code*, whose rules have been defined in advance
- An *encoder* with an encoding table
- A *digital message* that obeys the rules of the code
- A *decoder*, also with its own decoding table

All coding systems, like HTML, bar codes, postal codes, and Morse code—and the genetic code—fit this definition.

Is Sunlight Code? Are Snowflakes Code?

Many people ask me this, wondering if there might be an easy way around the "codes only come from intelligence" observation. The answer is no. That's because purely physical systems like rocks, snowflakes, and sunlight don't fit the definition (encoder—message—decoder). Rocks and snowflakes don't have encoders or decoders. Sunlight, for example, only becomes code when an eye or a sensor *encodes* the image as data, and then another device or your brain *decodes* the data.

---

* The word *obey* is appropriate in this context, because you can objectively determine whether the instructions were followed *correctly* or *incorrectly*. The cell monitors this and uses multiple checkpoints to halt the process if copying errors occur, including an "SOS" response to corrupted data. It uses sophisticated machinery to correct errors (659).

It is common in information systems parlance to say that *data* only becomes information after meaning is assigned to it. In DNA, the ribosomes assign meaning to the code in messenger RNA. The message in DNA is meaningful because the cell has a system for understanding it. In appendix 4 you will find a procedure for determining whether any system contains a code or not.

As I've stated elsewhere in the book, another reason why DNA is literally and not figuratively a code is that DNA has features only found in sophisticated codes: codes within codes (675), redundancy (326), checksums (316), and error correction (307). Hubert Yockey's book *Information Theory, Evolution, and the Origin of Life* (326) is a definitive text on this subject. (For a rigorous definition of codes, see appendix 4.) Again, everything in genetics, bioinformatics, and this book is based on this crucial fact.

When a cell reads instructions from DNA, an enzyme called a polymerase encodes data from the DNA into messenger RNA. Then the ribosomes decode it into proteins. The chaperones (special proteins that assist folding) organize these proteins into 3-D physical structures.

This decoding process is very similar to what happens when your friend prints out the Word document you just sent her. Your document is encoded into the USB cable connecting her computer to her printer, which decodes the 1's and 0's and prints the result on a physical piece of paper.

Cells Correct Errors

Whenever you have a code, external factors threaten your message. Error correction is essential. When I began studying it, I realized *error correction is absolutely central to the evolution question*. As a result of writing my Ethernet book, I knew it was not possible for innovations to come from copying errors. As I investigated what cells actually do, I discovered that not only do cells correct errors . . . cells also improvise when data has been lost! I'll circle back to this in chapter 11.

Meanwhile, this led me to a testable hypothesis: When DNA's code is fully unraveled, we'll find its data scheme safeguards more precious data with less wasted space than any existing human-made scheme. The error correction scheme for DNA is not well understood. DNA has vast regions of code that presently make little sense to geneticists. I suspected that much of this mysterious code is backup systems, error correction, and adaptive machinery.

Okay, So Where Does the Original Code Come From?

Messenger RNA is strings of code, generated by an encoder (RNA polymerase). This leaves us with the obvious mystery: Where did the original code in the DNA strand come from? Of course it came from the organism's parents, but where did the first cell get *its* code?

This led me to a crucial question: Is there any process in the natural world, outside of living things, that produces codes? Can you bridge the chasm between life and nonlife with anything other than intelligence? *This* was the chasm between Darwin and Design. Do nonliving things possess this kind of self-organization? And where do living things get *their* self-organization? How do they "know" how to respond to a threat?

This only evoked even more fundamental questions about self-organization, which we'll explore in future chapters:

- Do the laws of physics produce codes? If you provide the necessary chemicals, will natural laws do the rest?
- "Chaos theory" describes how nature produces sand dunes and hurricanes and tornadoes. Nature has all kinds of self-organizing properties. Does chaos also produce codes?
- Nobody has to design a snowflake or a tornado. Do codes ever emerge from chaos naturally, the way snowflakes do?
- Can complex chemical reactions produce codes?
- Is there any such thing as a naturally occurring code? Have we ever witnessed any brand new code coming into existence without someone having to design it?

James Watson shared the Nobel Prize for discovering DNA, the secret of life. He figured out how it's constructed and what chemicals it's made of. He proclaimed it contained no special substance or divine secret; just physics and chemistry.

But he never found out where code itself came from.

CHAPTER 8

Code First, Evolution Second

So these are the ropes,
the tricks of the trade,
the rules of the road.
You're one of the dopes
for whom they were made,
the rules of the road.

—NAT KING COLE

MY FRIEND AND COLLEAGUE Joel Runyon blogged this wonderful story about a chance encounter in a Portland, Oregon, coffee shop:

I sat down at yet another coffee shop in Portland determined to get some work done, catch up on some emails and write another blog post.

About 30 minutes into my working, an elderly gentleman at least 80 years old sat down next to me with a hot coffee and a pastry. I smiled at him and nodded and looked back at my computer as I continued to work.

"Do you like Apple?" as he gestured to the new MacBook Air I had picked up a few days prior.

"Yeah, I've been using them for a while." Wondering if I was going to get suckered into a Mac vs. PC debate in a Portland coffee shop with an elderly stranger.

"Do you program on them?"

"Well, I don't really know how to code, but I write quite a bit and spend a lot of time creating online projects and helping clients run their businesses."

"I've been against Macintosh company lately. They're trying to get everyone to use iPads and when people use iPads they end up just using technology to consume things instead of making things. With a computer you can make things. You can code, you can make things and create things that have never before existed and do things that have never been done before.

"That's the problem with a lot of people," he continued, "they don't try to do stuff that's never been done before, so they never do anything, but if they try to do it, they find out there's lots of things they can do that have never been done before."

I nodded my head in agreement and laughed to myself—thinking that would be something that I would say and the coincidence that out of all the people in the coffee shop I ended up talking to, it was this guy. What a way to open a conversation.

The old man turned back to his coffee, took a sip, and then looked back at me.

"In fact, I've done lots of things that haven't been done before," he said half-smiling.

Not sure if he was simply toying with me or not, my curiosity got the better of me.

"Oh, really? Like what types of things?" all the while, half-thinking he was going to make up something fairly non-impressive.

"I invented the first computer."

Um, Excuse me?

"I created the world's first internally programmable computer. It used to take up a space about as big as this whole room and my wife and I used to walk into it to program it."

"What's your name?" I asked, thinking that this guy is either another crazy homeless person in Portland or legitimately who he said he was.

"Russell Kirsch."

Sure enough, after .29 seconds, I found out he wasn't lying to my face. Russell Kirsch indeed invented the world's first internally programmable computer as well as a bunch of other things and definitely lives in Portland. As he talked, I began Googling him. He read my mind and volunteered:

"Here, I'll show you."

He stood up and directed me to a variety of websites and showed me through the archives of what he'd created while every once in a while dropping some minor detail like:

"I also created the first digital image. It was a photo of my son."

At this point, I had learned better than to call Russell's bluff, but sure enough, a few more Google searches showed that he did just that.

My colleague Joel Runyon accidentally bumps into the man who designed the first programmable computer and scanned the world's first digital picture, Russell Kirsch.

> Want to mess with your mind? Without the man in the photo, the photo of this man wouldn't exist. *mind blown*
>
> As he started showing me through the old history archives of what he did while any hope of productivity vacated my mind, I listened to his stories and picked his brain about what he had done.
>
> At some point in the conversation, I mentioned to him:
>
> "You know Russell, that's really impressive."
>
> He said, "I guess, I've always believed that nothing is withheld from us that we have conceived to do. Most people think the opposite—that all things are withheld from them which they have conceived to do and they end up doing nothing."
>
> "Wait," I said, pausing at his last sentence "What was that quote again?"
>
> "Nothing is withheld from us that we have conceived to do."
>
> "That's good, who said that?"
>
> "God did."
>
> "What?"
>
> "God said it and there were only two people who believed it, you know who?"
>
> "Nope, who?"
>
> "God and me, so I went out and did it."
>
> Well then, I thought—as he finished showing me through the archives—I'm not going to argue with the guy who invented the computer. After about 20 minutes of walking me through his contributions to technology, he sat down, finished his coffee, glanced at his half-eaten pastry, now-cold, checked his watch, and announced:
>
> "Well, I have to go now." (711)

How amazing that my friend Joel would bump into the guy who invented the first computer!

How Do You Invent a Computer?

How do you go from conceiving something to reality? How do you make the world's first digitized picture?

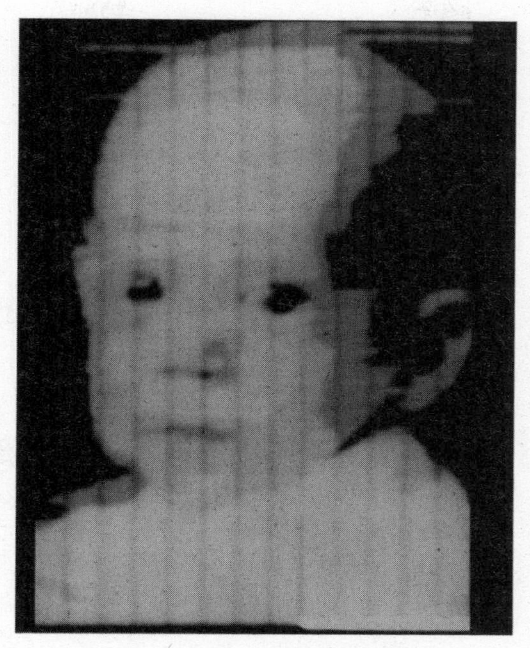

The world's first digitized photo, of Russell Kirsch's son Walden. Made in 1957, it's made of 30,976 black-or-white pixels.

Stop and think about what Russell Kirsch had to do to scan the first digital image. He had to decide that 1 means "on" and 0 means "off." He had to assign 1's as "white" and 0's as "black." He had to choose how many pixels wide the scanner was to be.

He had to decide how to store this data, what format it would be stored in, and how to retrieve it. He had to plan how the data was moved between registers within a very tiny amount of memory space. The printer had to render images in the exact same format they were scanned in. Each of these decisions required Kirsch to define many layers of language in detail *before a single circuit could be built.*

So now the question I needed answered was: How does a *cell* go from conceiving something to reality?

CHAPTER 9

Let's Make Some Noise About Noise: Dispelling Random Mutations Once and for All

Come on, feel the noise
Girls, rock your boys
We'll get wild, wild, wild

—QUIET RIOT

WHEN PEOPLE ASK ME how I met my wife Laura, I say, "I was the sound man and she was the singer. She fell in love with my reverb."

Being the man in charge of the decibels did snag me a girlfriend, but the life of a sound tech is no bed of roses. One evening when I was 17, I was the running sound for a concert with over 1,000 people in attendance. A choir was on stage, singing to a soundtrack on a cassette tape.

Gradually the soundtrack grew muddier and muddier. The beats were mixed up and the choir was getting lost. Altos and sopranos started drifting apart. For the life of me, I couldn't figure out what was going on. A sick sensation began to grow in my stomach.

Finally, the choir director motioned for the choir to stop. He turned around, staring at . . . *me*. Everyone else in the entire auditorium turned their heads and stared at me, too.

The cassette was playing two or three stretches of music simultaneously. Total cacophony. Now that the choir was no longer singing, I could tell the music from one layer had bled onto the next layer. Someone had put the cassette next to a magnet! Multiple parts of the same song were mixed together, eerily fading in and out with a watery, warbling tone. No wonder all the singers were lost.

I stopped the tape. My friend Jason sprinted up the aisle and out the door, frantically searching for a better copy of the soundtrack.

A stray magnetic field had turned our precious tape into garbled, useless noise.

Every recording engineer, every sound technician, every designer of radios, CD players, or cell phones wages a war on noise every day. Noise is your foe, whether you're trying to have a conversation in a crowded restaurant, or your car radio or cell phone is trying to pick up a signal. And yes, noise can humiliate you at the worst possible times.

Noise is anything that interferes with the signal, like stray magnetic fields. The sun is the number-one source of radio and TV noise—it generates that hiss you hear between stations on your TV or radio.

In information, noise—defined as random, unwanted disturbance to a wanted signal—is your enemy.

DNA is a code-based communication system. That means DNA battles noise, because all communication channels battle noise. Random DNA copying errors are nothing more than noise (314, 325, 308, 319, 300). Remember the fruit flies? The DNA copying errors induced by those experiments never seemed to produce anything good.

So I wondered: Are there ever situations where noise can work to your advantage?

I wasn't sure. But I knew how to find out. I had to study noise.

Claude Shannon's paper "A Mathematical Theory of Communication" was the natural starting point for this investigation. Shannon's work, written in 1948 (320), made possible our current digital era by pioneering information theory; *Scientific American* called it "the Magna Carta of the Information Age." Nassim Nicholas Taleb, author of *Antifragile* and *The Black Swan*, called information theory "the Mother discipline."*

A 1953 issue of *Fortune* said, "It may be no exaggeration to say that man's progress in peace, and security in war, depend more on fruitful applications of Information Theory than on physical demonstrations, either in bombs or in power plants, that Einstein's famous equation works." Since your cell phone, computer, and video player all critically depend on Shannon's theory, information theory is one of the most practical mathematical discoveries of the 20th century.

Shannon's paper defined how information is encoded, transmitted, and decoded. His most dramatic discovery was a remarkable parallel: The math that describes the uncertainty of noise is exactly the same as thermodynamic entropy. He called noise "information entropy."†

Information entropy is not nearly as well known as heat entropy, so allow me to explain both.

To most people, "entropy" is the tendency for their teenage daughter's room to descend into utter chaos ("Hey, wasn't this place almost perfect just two days ago?"), and that definition is pretty accurate. Formally, heat entropy is the principle in thermodynamics that says that the path from order to disorder is irreversible (253). Entropy is partly why, after you've burned a tank of gas in your car, the exhaust is never going to rush back into your tailpipe, reverse-combust in your engine, and turn back into gasoline.

Information entropy is similar, but it applies to data instead of heat. It's not just a pattern of disorder; it's also a number that measures the loss of information in a transmitted signal or message. A unit of information—the bit—measures the number of choices (the number of possible messages) symbols can carry. For example, an ASCII string of seven bits has 2^7 or 128 message choices. Information entropy measures the uncertainty

---

* Taleb says, "Born in, of all disciplines, Electrical Engineering, the field has progressively infiltrated probability theory, computer science, statistical physics, data science, gambling theory, ruin problems, complexity, even how one deals with knowledge, epistemology. It defines noise/signal, order/disorder, etc. It studies cellular automata. You can use it in theology (FREE WILL & algorithmic complexity). As I said, it is the MOTHER discipline. I am certain much of Medicine will naturally grow to be a subset of it, both operationally, and in studying how the human body works: the latter is an information machine. Same with linguistics. Same with political "science," same with... everything" (www.facebook.com/13012333374/photos/a.10150109720973375.279515.13012333374/10152488919783375/?type=1).

† As an indication of the relevance of Claude Shannon's paper to genetics, go to http://scholar.google.com and search for "Shannon 1948 genetic code." At the time of this writing, Google returns more than 8,000 books and papers.

that noise imposed upon the original signal. Once you know how much information you started with, entropy tells you how much you've lost and can't get back.

Information entropy is not reversible, because once a bit has been lost and becomes a question mark, it's impossible to get it back. Worse yet, the decoder doesn't report the question mark! It assigns it a 1 or 0. Half the time it will be right. But half the time it will be wrong, and you can't know which bits are the originals and which bits are only guesses.

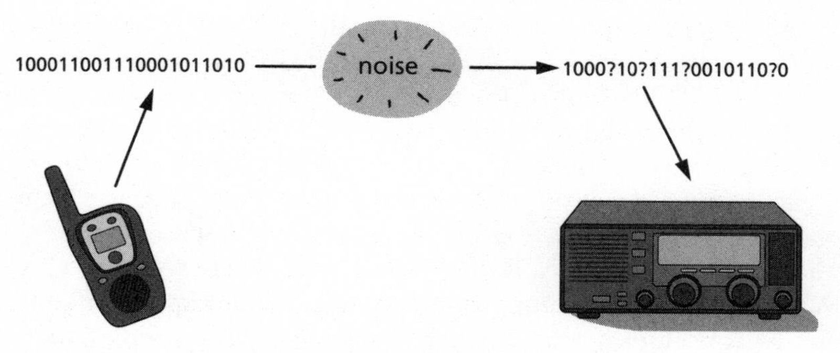

The question marks in the received signal are bits that have become lost in noise.

Noise equals uncertainty. When there's no noise and you receive a 1, you are 100 percent sure the transmitter sent a 1. The more noise there is, the less certain you are what the original signal was. If there's lots of noise, you're only 50 percent sure the transmitter sent a 1, because your decoder gets it wrong half the time. Nobody knows what was originally said. (In chapter 19 you'll see that cells encode and decode information constantly, according to linguistic rules [403].)

Because signals (encoded messages) and noise are polar opposites, coded information can never come *from* noise. A broken machine makes a horrible squeal. But there's no encoder, so the squeal is not code. (It's not digital either. It's an analog sound wave.) An intelligent agent has to encode that squeal into digital symbols and interpret their meaning before it can be considered information.

In my search for books that explained how information theory relates to DNA, I came across one by Werner Gitt, called *In the Beginning Was Information*, written in 1997 (310).

Gitt was a Young Earth Creationist, which raised my hackles (remember, I had decided against a young Earth as there was too much evidence to the contrary). But as it turned out, very little of his book was about his Young Earth Creationism beliefs. It was a rigorous treatment of digital data by a skilled engineer. His elegant approach to information in biology was precise and compelling.

Gitt laid out a hierarchical structure for DNA, and he showed the aforementioned fact that information never comes from noise,* elaborating upon the work of Claude Shannon. Even more important, Gitt pointed out that all communication systems with digital code that match Shannon's model and whose origins we know are *designed.*

There's a very simple explanation for this. The alphabet (symbols), syntax (grammar), and semantics (meaning) of any communication system must be determined in advance, *before* any communication can take place. Otherwise, you could never be certain that what the transmitter is saying is the same as what the receiver is hearing. It's like when you visit a Russian website and your browser doesn't have the language plug-in for Russian. The text just appears as a bunch of squares. You would never have any idea if the Russian words were spelled right.

When a message's meaning is not yet decided, it requires intentional action by conscious agents to reach a consensus. The simple process of creating a new word in English, like *blog* (which was originally *web log*), requires speakers who agree on the meaning of the other words in their sentences. Then they have to mutually agree to define the new word in a specific way. Once a word is agreed upon, it is added to the dictionary. The dictionary is a decode table for the English language.

Even if noise might occasionally give you a real word by accident, it could never also tell you what that word means. Every word has to be defined by mutual agreement and used in correct context in order to have meaning.

* Gitt is criticized by skeptics, but so far as I've been able to tell, their assertions are unfounded. Gitt's analysis is entirely consistent with Claude Shannon's definitions. In every case I could find, Gitt's critics abused or misused Shannon's work on major points. A good example of a criticism based on a fundamental misunderstanding of Shannon can be found on the website TalkOrigins, discussed in detail at www.cosmicfingerprints.com/talkorigins-gitt.

Loss Information Is Like Hot Toast Grown Cold

To better understand why you can't just reverse the loss once information is lost, think about toast.

You toast bread in your toaster and it heats up. Then, when you take your toast out of the toaster and put it on a cool plate, the heat radiates away from the toast and into the cold room.

Heat entropy refers to energy loss—here, your toast growing cold. This loss of heat is irreversible (unless you put it back into a system/toaster, which increases its heat again). It will never get hot again all by itself—ever. The laws of thermodynamics dictate that entropy (loss of useful energy) always increases in a closed system.

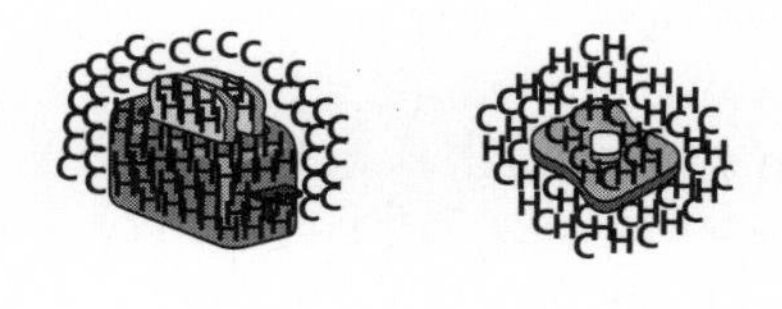

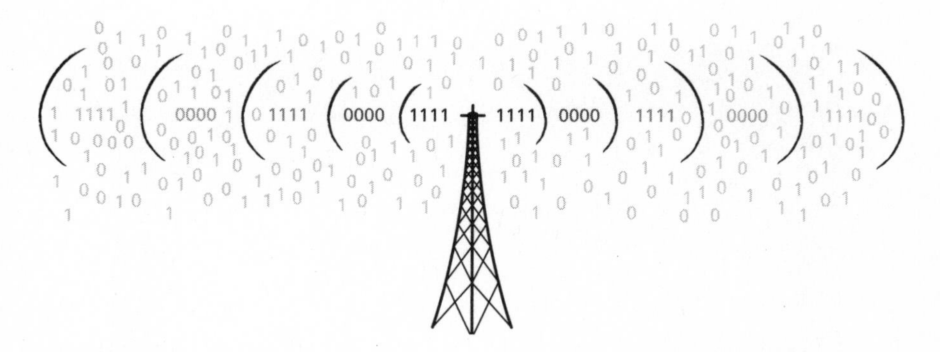

The math for toast getting cold is identical to the math for signals getting noisy as you travel away from an antenna. Thermodynamic entropy and information entropy are both irreversible. Cold toast never gets hot, and noise never adds useful content to a signal.

Let's look at this a different way.

When CDs were new, stereo shops would demonstrate the quality of their new cassette decks by switching back and forth between the CD and the taped copy. A good cassette came surprisingly close to the CD in sound quality, but it definitely was not perfect. Recordings had obvious "tape hiss" between songs.

The tape adds noise. It's on the tape before you even press the record button. Once it's there, you never get rid of it. You can use Dolby™ noise reduction but it doesn't get rid of the noise; it just hides it. Everyone who's watched a snowy TV show from a station 60 miles away knows you can smack the antenna around and reduce the noise, but you can never convert a snowy TV picture back to a clear TV picture after it's become noisy.

If 100,000 homes pick up that TV signal 60 miles away, each TV receives a slightly different version of the noisy, snowy picture. Each home has lost different bits of the signal. And every single copy is inferior to the original.

Random Mutation = Noise

Every communication system battles noise. Noise is a random change of a signal—in other words, noise is a *mutation* of the original message. In a TV signal, random mutation looks like this:

You can listen to it, too. Just tune your radio dial to the space between stations and listen to the hiss.* Or listen to the sound of air whooshing out of a vent in the ceiling. *White noise.*

When you want to watch your favorite TV show, noise is your enemy. Once noise is with you, it's with you forever. Toast never gets hotter after you take it out of the toaster. CDs never sound better after you copy them onto a cassette tape; you can make a million copies of that tape, and even if natural selection cooperates perfectly, the best version you can get is to select the least inferior version.

Sometimes people have disagreed, offering hypothetical scenarios where a few random changes might just happen to give you the exact step you need. However, since a random change could happen anywhere across a billion base pairs in the genome, getting two or three or five errors to occur in an advantageous way without breaking something else first is statistically all but impossible. Again, nobody appreciates this more than a communications engineer.†

Random mutation is noise (317, 326) and noise destroys. Random mutations = damaged DNA (238). We saw that in the fruit fly experiments.

Any randomness-based theory of evolution violates the laws of information entropy. Music doesn't get better when you scratch CDs. Organisms do not gain new features when their DNA mutates through damage or copying errors. Instead they get cystic fibrosis or some other birth defect, like legs instead of antennae growing out of a fruit fly's head. Natural selection can clear competition by killing off inferior rivals. But it can't work backward from a random mutation and undo the damage.

Biologist Lynn Margulis, wife of Carl Sagan, was a professor at the University of Massachusetts at Amherst. She espoused the theory of Symbiogenesis, which says that cells form interdependent relationships with other cells in order to create new features. We'll come to this in chapter 15.

---

* It's not possible to prove that noise from the sun—the interference that creates white noise—is random. If we knew every last detail of every subatomic particle, maybe we could predict it. However, unlike DNA's adaptive mutation systems, which I discuss in chapters 11–16, it does pass all the statistical "smell tests" for randomness (e.g., a spectrum analysis to check if white noise is really "white"). Whether it is absolutely random or not, the sun's radiation is at least noise with respect to your radio or TV.

† In a few categories of special situations, noise can also be helpful. When it's applied intentionally, noise is called *dither*. I explain dither in appendix 1.

Margulis was well aware of those failed fruit fly experiments. She observed, "Many ways to induce mutations are known but none lead to new organisms. Mutation accumulation does not lead to new species or even to new organs or new tissues ... Even professional evolutionary biologists are hard put to find mutations, experimentally induced or spontaneous, that lead in a positive way to evolutionary change" (637).

For many decades, the Neo-Darwinian Modern Synthesis has claimed that adding noise to a signal can occasionally improve its content. Beneficial random mutations, together with natural selection, were allegedly the key to everything. If this were actually the case, I would have to agree that Mother Nature would possess a truly amazing built-in tool of continuous improvement.

How intriguing it was, then, to confirm that in computer science, networking, and telecommunications, the concept of adding noise to a signal to improve its contents simply *does not exist at all*; neither in theory nor practice. Claude Shannon's work showed the exact opposite. This confirmed my suspicions about the fruit fly experiments: They were doomed to fail from the start.

Much as some might like chance and selection to be all you need for evolution, the real world appears to demand far more powerful approaches than that.

BULLET POINT SUMMARY:

- Neo-Darwinism says Random Mutation + Natural Selection + Time = Evolution.
- Random Mutation is noise. Noise destroys.

CHAPTER 10

How Do We Fix Evolution?

I'm in repair
I'm not together but I'm getting there
I'm in repair
I'm not together but I'm getting there

—JOHN MAYER

IN 2004, the School District of Dover, Pennsylvania, dictated that biology teachers must present Intelligent Design as an alternative to evolution. The board ruled that teachers must read a statement aloud in ninth-grade science classes when evolution was taught. The statement described ID as "an explanation of the origin of life that differs from Darwin's view."

Eleven parents became so angry, they sued the Dover Area School District. The American Civil Liberties Union and the Darwin lobby jumped in and supported them. The court case became an international news sensation overnight.

Pro-evolution lawyers and witnesses successfully argued that ID is a form of Creationism. Judge John E. Jones ruled that ID was not science because a divine Designer cannot be tested in the lab.

In just a few chapters I'll address the question of whether Intelligent Design is scientific or not, because I'm offering a large technology prize for the answer. So hang tight.

Meanwhile . . . you know what the irony is?

Neo-Darwinism isn't a scientific theory, either!

Darwinism can't be tested in the lab. It can't be proven with mathematical models.

Here's why.

Darwinism claims that random mutation combined with natural selection over time causes evolution. However, *the claim that any mutation is random is not provable. This is because, in mathematics, there is no method for proving any sequence of numbers or symbols is random. In fact, to prove that anything is random has been proven impossible* (800)!

No test in mathematics or science can confirm that a mutation is random; the book *Information, Randomness & Incompleteness* by Gregory Chaitin explains why. Therefore it is impossible for science to demonstrate that the Modern Synthesis is true. (Appendix 1, "All About Randomness," unpacks this problem in detail. If you're not sure I'm telling you the truth, jump over to appendix 1. Also be sure and check the many excellent references.)

You can't prove the letters in any word are random. You can't prove that the words in any sentence are random; you can't prove the sentences in any paragraph are random. You can't prove the base pairs in any gene are random. It's impossible.

The fact that noise destroys explained why Dobzhansky and Goldschmidt's fruit flies never evolved. Of course they didn't evolve; they were doing everything they could to resist *de*volving!

There is no such thing as a percentage of time that a corrupted signal is "better." It's always worse. How can natural selection compensate for the loss if every mutant is inferior to begin with?

This means that DNA copying errors trigger a species' decline (246), and natural selection finishes it off. To revise our evolution formula, then:

RANDOM MUTATION + NATURAL SELECTION + TIME = <u>EXTINCTION</u> NOT EVOLUTION

Radiation didn't make fruit flies evolve. *But this still didn't tell me whether or not evolution was true.* Some people (e.g., Werner Gitt) had reached this point and stopped, declaring evolution a hoax. But as you're about to find out, what I've explained to you so far is barely the tip of the iceberg.

While this disproved the theory that evolution was the result of random mutations, none of it eliminated the possibility of evolution at all! Instead, it revealed an utterly tantalizing mystery. Could it suggest that the real mechanisms of evolution are vastly more fascinating and ingenious than we ever imagined? I was intrigued, since a "nonrandom mutations" theory of evolution would be a scientific theory because scientists and mathematicians can easily check for nonrandom changes in DNA patterns.

Not only that, but if genetic information has been preserved since the earliest life forms more than 3 billion years ago, DNA must possess some kind of incredibly powerful error correction system.

Yes, there is surely some vanishingly small number of beneficial mutations that were generated by random accidental copying errors. But there's no way to be certain they were random. Experimentally, they're as rare as blasting fruit flies with radiation and getting a new species.

In other words, one scientifically untestable theory ("an intelligent designer created new species") has been duking it out with another scientifically untestable theory ("the mutations that create new species are random and accidental"). It is impossible for science to validate either assertion. So our deadlock between Darwin and Design shouldn't be all that surprising.

Random mutations are philosophical and metaphysical assertions, not provable scientific theory.* *Random*, by definition, means further explanation as to how they got that way is not possible. So the minute someone insists any process in nature is random, scientific inquiry stops cold. This is why "random mutations" deserve no place in modern theories of evolution. (Again, see appendix 1, "All About Randomness," which dismantles the random mutation hypothesis.)

How do we pry evolution out of its antiscientific straitjacket of randomness?

Only by proposing that the mutations aren't random, but rather follow some sort of formula or pattern. Then and only then can we have a properly scientific theory of evolution.

---

* Every philosopher knows science cannot prove God. Hopefully most scientists know this, too. But science's intrinsic limitations don't mean God doesn't exist. Therefore, "what is scientifically testable and what is not" by definition can never be an ultimate measure of truth. As Kurt Gödel (800) proved in 1931, no system of logic can prove its own assumptions. Since science always relies on philosophical and metaphysical assumptions that lie outside of science, it is intellectually dishonest to ban debates about God and origins from the classroom just because God cannot be scientifically tested. It is possible to admit the fact that science (information theory in particular) infers a designer without endorsing a specific religion. And as the Evolution 2.0 Prize described in chapter 23 and appendix 4 attests, we must always remain open to the fact that science may resolve currently unanswered questions.

There are many ways to prove mutations are nonrandom. If you see the word *and* 1,000 times in a 50-page book, you can be sure it did not get there by accident. It appears in specific places and follows specific rules. If you can show that certain patterns appear again and again (such as mice and humans sharing 99 percent of the same genes; see ref. 322), then you have definitive *proof* that the sequence is *not* random.

Stay tuned, because scientific research has revealed classes of mutations that are not random at all—observable, adaptive mutations, which happen literally every day. A scientifically testable version of evolution becomes possible!

Can you improve a signal by changing it? Sure you can, as long as you change it within the rules of the language. You can edit a Microsoft Word document and add new content. You can't edit the doc with an arc welder. And you have to add real English content to it.

Does this mean that . . .
Living things adapt because cells edit their own DNA?

That idea made sense to me. I liked it, because a theory could be tested in the lab. In my studies, I found that fewer and fewer biologists in the 21st century hold to a "pure randomness" concept of mutations anymore.

Might this mean that the source of evolution could be . . . the decision-making ability of the cell itself?

If so . . . then codes and error correction necessarily *preceded* DNA and the first cell.

As you're about to discover in the very next chapter, evolution is possible only because cells are goal seeking and actively edit their genomes in response to threats (643, 645). Many times, they make excellent choices.

The current explanation for evolution—randomness—is wrong. We know this from information entropy; we know this from fruit fly experiments. So . . . what is the real explanation?

What I found was a "Swiss Army Knife" with five blades, and a new way of thinking about evolution, a more advanced way—call it Evolution 2.0.

HOW EVOLUTION *REALLY* WORKS

CHAPTER 11

Blade #1: Transposition—The 70-Year-Old Nobel Prize–Winning Discovery Nobody Talks About

A secret, a secret, he says I've got a little secret,
A secret, a secret, a secret kind of secret.
I'm aching for to shout it to every daffodil
And tell the world about it, in fact I think I will.

—DEAN MARTIN

PREGNANT WOMEN who arrived at Vienna General Hospital's Obstetrics Clinic One in 1846 begged to be admitted to Obstetrics Clinic Two. Clinic One was well known for maternal deaths from childbed fever. Sometimes these women pleaded with the admissions staff on their knees.

That year, the death rate from childbed fever in Clinic One was 11.4 percent, 260 deaths total. But in Clinic Two it was only 2.7 percent.

Some women, upon being refused admission to the second clinic, turned and walked out the door, choosing to give birth in the open streets. They knew their chances of an infection-free birth and escaping death were better on the street than in the Hospital.

Dr. Ignaz Semmelweis resolved to solve this riddle by conducting a series of experiments. One such experiment concerned the priest bringing last rites to dying women. He passed through five wards before reaching the final sickroom.

When patients would hear the priest's attendant and her ringing bell, they would shudder with fright. The women believed this increased their chances of succumbing to the fever. Semmelweis persuaded the priest to walk a different way without the ringing bell, but deaths continued.

Another suggestion was that women in Clinic One delivered babies while lying on their backs, while women in Clinic Two were on their sides. Semmelweis tested that, too. Again, no difference.

The following year, Semmelweis' colleague Jakob Kolletschka cut his finger with a scalpel while he was performing an autopsy. Autopsies were routinely performed in one section of Clinic One. Jakob died in agony with symptoms of childbed fever.

Semmelweis theorized that "cadaveric matter" from the scalpel had gotten into Kolletschka's bloodstream. He proposed that the medical staff and students became carriers of infectious material.

He noticed that the smell of cadavers was still strong on doctors' hands after they washed and went into the maternity ward. He also noticed that if he washed his hands in chlorinated lime, the smell disappeared.

He ordered all medical students to wash their hands in his new disinfectant, and in 1848, mortality from childbed fever dropped to 1.27 percent in Clinic One. Mortality in Clinic Two was 1.33 percent. During two months in 1848, the death rate was zero.

At this time, no one understood the true nature of bacteria and pathogens. Back then, scientists ascribed disease to "atmospheric-cosmic-telluric changes" across entire regions. Some had theories about overcrowding, but Clinic Two was more crowded than Clinic One.

Despite Semmelweis' success, his colleagues and the medical profession in general refused to accept his findings. In absence of hard evidence of these "cadaverous particles," critics scorned his theory as magical and superstitious. Did Semmelweis really expect educated men to believe that "corpse particles" might turn a person into a corpse, with no specified mechanism, after a single contact? Ridiculous.

Semmelweis was declined reappointment at the Vienna hospital in 1849. He moved to Hungary, where he took a position in a maternity ward. There, mortality rates also fell dramatically.

Word of Semmelweis' success spread across Europe...and so did opposition to his ideas. Dr. Charles Meigs, a leading obstetrician from Philadelphia, retorted, "Doctors are gentlemen, and gentlemen's hands are clean." The head of the Copenhagen maternity hospital argued, "It seems improbable that enough infective matter or vapor could be secluded around the fingernails to kill a patient."

Semmelweis published a book on his results and became obsessed with childbed fever. He drank heavily. He began lashing out at his critics in a series of open letters, denouncing them as irresponsible "murderers" and "ignoramuses."

In 1865, suffering from stress, overwork, and possible dementia, he was referred to a mental institution. He was lured into the asylum under the pretense of visiting a colleague's institute. When he realized what was happening, he struggled to leave. He was seized by guards and severely beaten, confined to a straitjacket, and thrown into a cell.

Two weeks later, he perished in a delirium of fever and chills, from a gangrenous wound inflicted by the guards. Ironically, the cause of his death was...infectious disease entering his bloodstream through open skin. Just like the colleague who had cut his finger performing an autopsy in Vienna.

A new director took his place at the Hungarian maternity ward. Hand washing became a thing of the past. Mortality rates skyrocketed 600 percent; thousands of women and babies died from the fever. No one objected.

Hand washing did not come into vogue until 20 years later, when Louis Pasteur popularized his germ theory of disease (251).

The term *Semmelweis reflex* refers to new knowledge being rejected because it overturns entrenched norms, popular beliefs, and accepted paradigms.

Barbara McClintock: Champion of the Smart Cell

A full century after Semmelweis, cytogeneticist Barbara McClintock performed meticulous experiments manipulating chromosomes in maize. Maize is known as Indian corn and has variously colored kernels.

DNA was still poorly understood in the 1940s, but scientists clearly understood that each cell contained genetic material with sections called genes and chromosomes. McClintock could observe chromosomes in her microscope. She would damage chromosomes and observed what happened based on the changes in the colored kernels of corn (639).

Dr. Barbara McClintock developed such intimate familiarity with maize that she could detect rearrangements of genes and chromosomes by studying the colored kernel patterns. Even though she had a reputation for her "intuitive" grasp of the plants she studied, in reality she painstakingly documented hundreds of plants and thousands of kernels in each experiment.

Remember the fruit fly experiments? McClintock's experiments were similar. She too used organisms damaged by radiation. She discovered that radiation broke chromosomes and triggered editing systems *in real time*. Cells would reconstruct the damaged chromosome with another section of radiation-broken genetic material.

Dr. McClintock began to construct a picture of what happened when she damaged a chromosome. In one experiment, she "hacked" her corn plants in a very clever and original way, almost like a modern computer programmer hacks into a system to reveal its vulnerabilities. Using plants that needed to align chromosome pairs with inverted code segments, she created a situation where cells, in their offspring, were forced to join broken chromosomes together repeatedly to reproduce successfully. Each generation created new instabilities and code combinations. These required further repairs for the plant to grow.

The plants she hacked were unable to recover their original information. A few of them improvised. They sensed the damage they could not repair and activated previously latent parts of the maize genome (640). They succeeded in patching damaged DNA with a *new* kind of genetic element—a *transposable* element. Their movement is called *Transposition*. Transposition is part of the cell's toolbox for re-engineering its own DNA. (See page 87.) By moving to new places, they changed expression of the genome. McClintock named them "controlling elements."

McClintock was a hacker in the noblest sense of the word. She subjected a system to something that was radically unexpected, and got a surprising and gratifying result. It was so surprising, it took decades for the world to see it.

The Intuition and Discipline of Barbara McClintock

When asked about the difference between the fruit fly experiments and McClintock's maize experiments, her colleague, Dr. James Shapiro of the University of Chicago, explained it this way:

> X-rays break chromosomes, triggering a built-in repair system that is used as a normal feature of life. So one would not expect much in the way of evolutionary innovation from X-ray exposure, although it does lead to chromosome rearrangements.
>
> McClintock posed her plants an entirely different type of challenge, which resulted in the activation of transposable elements and other genome restructuring activities. She explains this, although in rather technical terms, in her Nobel Prize speech (640). Basically, she gave them a single broken end that could not be joined to another end until the chromosome had duplicated; the result of joining the two duplicated ends was a chromosome with two centromeres* that would go to opposite daughter cells at division, creating a chromosome "bridge" that had to break for division to complete. This "breakage-fusion-bridge" cycle was a continuous genome instability that had to be resolved for normal growth to resume.
>
> However, her hallmark was her character and personality. McClintock was a very special person. I tried to capture this in my obituary of her (657). She had an exceptionally curious and open mind and paid close

* A centromere is the part of a chromosome that links chromatids, which are copies of a duplicated chromosome. Think of it as a shipping label that tells the cell where to send the chromosome.

attention to what the plants were telling her. She also had a keen sense of the temporary nature of fashionable scientific notions, as she also explained in her Nobel Prize address. Combined with a furious work ethic and deep knowledge of maize cytogenetics and how to exploit them experimentally, these traits contributed to her amazing accomplishments. (Private correspondence with James A. Shapiro, May 2, 2012)

In other words, even though noise always destroys information, the plot had thickened. Barbara McClintock had discovered that plants possess the ability to recognize that data has been corrupted. Then they repair it with newly activated genome elements, and in the process of repairing the data, the plants can develop new features!*

McClintock did not have to kill her plants to get them to adapt; she only had to damage a chromosome—meaning that natural selection was not essential to the evolutionary process.

At least in some cases, nothing died. (Of course she also provided the best possible conditions for these plants to flourish.) Yet the plant was still evolving. Why? Because it was actively working to safeguard its genome for the next generation. The plants then passed these activated transposable elements to their offspring.

McClintock showed that you can get variation and adaptation *before* natural selection even has a chance to do its culling.

Semmelweis Reflex to Transposition

In 1951, McClintock presented her findings to a symposium at Cold Spring Harbor in New York. Biographer Evelyn Fox Keller (627) wrote that the audience's reaction in 1951 to Transposition was "stony silence," punctuated only with muttering and muffled laughter. She described the reception of her research as "puzzlement, even hostility." She experienced what Dr. Semmelweis had experienced almost exactly 100 years before: accusations of mysticism and superstition.

* Geneticist Evelyn Witkin did similar experiments with bacteria, also in the 1940s, causing them to mutate with UV light. She said, "It provided us with a system of quantifying induced mutations and of seeing repair before our eyes, almost" (309). Unlike the fruit fly experimenters, Witkin and McClintock, close friends, were sensitive to the fact that organisms were actively repairing their genomes.

You might wonder if she got this reaction simply because she was a woman. While she certainly was a pioneer in that regard, she had earned considerable respect and was generally treated well (642). But her audience had no grid for anything like . . . *this*.

Unlike Semmelweis, McClintock didn't flame her critics with nasty letters. She didn't start drinking, either. Instead she stepped up her commitment to research and resolved to not broadcast her work. For the next 20 years she was pretty much quiet about her discoveries. (How rare it is for people far ahead of their time to stick to their guns and patiently wait for history to prove them right!)

Then, in 1968, McClintock's colleague James Shapiro confirmed bacteria could also transpose elements in DNA (600). In the 1970s Transposition began to receive wide recognition as a vital mechanism of genome change. More researchers independently confirmed what McClintock had already discovered: Organisms edit their DNA.

In 1983 she received the Nobel Prize for her discovery of Transposition. In 2005 her picture appeared on a U.S. postage stamp, which includes a diagram that shows how she set up cells to re-arrange DNA segments. She became a science celebrity. Scientists around the world enlisted her help to solve seemingly intractable problems. Speaking invitations poured in from all over the world, and Harvard University granted her an honorary PhD. Today she is regarded as one of the finest scientific minds of the 20th century.

The Barbara McClintock postage stamp illustrates her greatest discovery, Transposition: Cells re-arrange chromosomes, pair them in new combinations, and insert them into another part of the genome.

So...why is Transposition omitted from entry-level biology classes even today? Why do the most popular evolution books neglect to say anything about this powerful, adaptive mutation system? Several popular textbooks (e.g., *Human Biology*, 7th ed., by Daniel D. Chiras) make no mention. Little or nothing is said about it in mainstream evolution books by Dawkins and Coyne.

McClintock's results are not omitted from mainstream evolution because her findings haven't been verified and accepted by the scientific community. They have, and any high school student is quite capable of understanding the concept. A sketch of the starting point of her discovery fits on a postage stamp!

Transposition is a poster child for the chasm between "evolution as fed to the general public" and "evolution as practiced by real biologists." All serious biologists know about Transposition. No one disputes it. But while you can easily explain it to a 10-year-old, my experience is that not one regular person in 100 has ever heard a thing about it. McClintock's work continues to get the Semmelweis treatment even now.

When you ask a biologist how evolution works, he'll usually say, "Random mutation and natural selection." But, as we've seen, the random mutation theory is wrong. And since nearly every cell in existence is capable of Transposition, a far more accurate answer would be "Transposition and natural selection." How come they're not telling us about Transposition?

All I can think of is that McClintock's discoveries fly in the face of old-school Darwinian dogma. Her mobile genetic elements were *anything but* random. Cells cut and splice their DNA in specific locations and patterns. Like Semmelweis' critics, people schooled in the Darwinian paradigm are blind to the import of McClintock's work.

In personal conversations on my blog I've found that many who believe evolution is random and purposeless deeply resent Transposition, because it implies that life follows some sort of plan or formula—precisely the notion that Darwin allegedly overturned in 1859. Again, as I said in chapter 10, any theory that is not searching for a pattern or formula behind DNA mutations is not science.

Transposition is "blade #1" of the Evolution 2.0 Swiss Army Knife. It's a central process for ongoing genetic innovation. Cells swap sections of their DNA when they need to adapt to their environment.

Think of Transposition as cut/copy/paste for genetic information inside a cell—like when you write an article and decide to rearrange paragraphs, or even pull out half of one chapter and reinsert it after a different chapter. Bacteria do this; plant and animal cells do it, too.

TRANSPOSITION

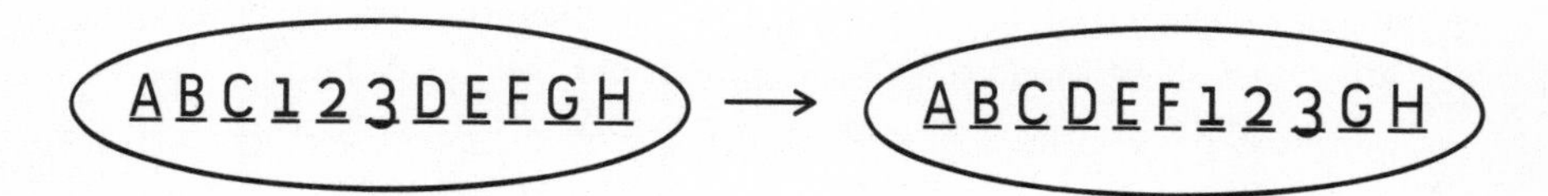

In concept, Transposition is very simple: re-arranging coding sequences A, B, C, etc., within DNA. In practice, it's dazzlingly sophisticated, just like a computer program that rearranges blocks of its own code. Protozoans are known to be able to rearrange 100,000 segments of their own DNA in real time.

When cells encounter hostile chemicals and threats, like McClintock's corn plant hacks, they adapt. It might even mean that when giraffes needed longer necks to reach the leaves high in the trees, they didn't have to wait for a random occurrence to make the change, but rather a system switched on and engineered a solution. McClintock's work implies that, like a computer program that rewrites itself on the fly, cells use their "mutation algorithm" to make smart substitutions—and a longer neck, like differently patterned corn, could be the result.

Old-School Darwinism Versus Evolution 2.0

Does Transposition contradict Neo-Darwinism? If we're going to use consistent terminology, yes, it does. Neo-Darwinism by definition says evolutionary changes are caused by random mutations and genetic drift. It emphasizes that these changes are gradual and are not goal driven.

Transposition and the other mechanisms that follow are targeted adaptations to threats. They are neither random nor gradual, because they respond to specific threats and exhibit known patterns. Genetic engineers get predictable results because they have learned they can manipulate Transposition in repeatable ways.

Even though many scientists have simply added Transposition to our understanding of evolution and labeled it "Darwinian," McClintock's work is a "Post-Darwinian" theory of evolution. I believe this is why her discoveries were resisted for decades, and why entry-level biology books say so little about it.

Clearly, theories of evolution have come a long way and have incorporated many new discoveries. To put any of those major discoveries of the last 50 years under the "Darwinian" umbrella is to constantly redefine Darwinism, giving Darwin far too much credit. Modern research shows that a great deal of the Modern Synthesis is obsolete. (643, 645, 646)

"You Mean . . . It's Not Random?"

James Shapiro, who discovered Transposition in bacteria, was not only a close friend of McClintock and carried on her work; he also holds an Order of the British Empire medal from the queen of England, and won the Darwin Prize Visiting Professorship from the University of Edinburgh. He was elected as a fellow of the American Association for the Advancement of Science for "innovative and creative interpretations of bacterial genetics and growth."

In 2010, I attended a lecture by Dr. Shapiro at Fermilab. That evening he gave a riveting talk on evolution. He described McClintock's career and numerous mechanisms like Transposition.

Shapiro described life's adaptive formula as a "hierarchical operation of cellular control regimes" (665), and the cell itself as "systems all the way down" (666). He continued: "There are piecemeal coding sequences, expression signals, splicing signals, regulatory signals, epigenetic formatting signals, and many other 'DNA elements'... that participate in the multiple functions involved in genome expression, replication, transmission, repair and evolution."

After his talk, a smaller group huddled around him in the Fermilab cafeteria, peppering him with questions. Suddenly one guy "got" what Dr. Shapiro had been saying all night long.

"You mean the mutations aren't random?" he asked.

"No sir," replied Dr. Shapiro, "they're not random at all. When bacteria are comfortable, some mutations cannot be found in over ten billion cells. But when they're starving, the mutation frequency can go by a factor of >100,000-fold and they develop new adaptations so they can survive" (658).

Cells are capable of doing their own genetic engineering—a natural version of what scientists do in experiments in labs.

Natural selection is not the driving force of Evolution 2.0. Natural genetic engineering is. This is defined as "the collection of the regulated biochemical functions a living cell possesses to restructure its genome" (659).

I watched the face of the man questioning Dr. Shapiro. You could almost hear the gears grinding inside his head! The man looked to be in his fifties; I imagine he'd bought into the random mutation myth decades ago and had never questioned it since.

Introducing . . . the Profound Protozoan

Research at the University of Colorado reports a fascinating discovery about protozoa. This is profound, almost miraculous:

Starving male and female protozoans mate (pooling their poverty, apparently), and then completely restructure the genome to make a new nucleus, cutting DNA into 100,000 pieces, then splicing and rearranging the code (650).

In other words, a protozoan reprograms its own DNA through a repeating, programmed response to stress—through thousands of simultaneous edits (613). They do this in response to heat shock, pollution, hazardous chemicals, absence of food, and presence of food they're unable to digest (659). They do this in a few hours!

A cell editing its own DNA is like a writer with 20 Microsoft Word documents open all at the same time, rapidly shuffling pictures, words, and tables back and forth. That's what's happening when a protozoan splices its own DNA into 100,000 pieces and rearranges them. Linguistic analysis of the protozoan genome shows that only certain segments get cut and spliced, and apparently only certain arrangements are allowed.*

* Natural Genetic Engineering is neither random nor 100 percent predictable ("deterministic"). It's *ergodic* (never entirely predictable but always obeying familiar patterns), a concept I elaborate on in appendix 1, "All About Randomness." Please do *not* assume that cells always successfully predict what changes they need to make! Cells make mistakes just like we do. But their success rate is far too high and the results too fast to ascribe to chance. We stand to learn a great deal from understanding the patterns they use when they modify their genomes.

Transposition is not only for bacteria and corn. A recent comparison of primate and human genomes (630) shows that mammals with placentas, including humans, share more than 280,000 transposable elements. Ever-expanding libraries of genome sequences infer that roughly 20 percent of differences between humans, primates, and related mammals vs. marsupials can be traced to massive rearrangements of blocks of data, via Transposition.

Non-Darwinian Evolution?

I began to wonder how it was even possible that biology books would say random copying errors generate the raw material for evolution, which is patently false—especially when a more sensible, easy-to-understand, proven alternative had already been well known to PhD candidates for half a century.

What an unfathomable disservice had been done to students and researchers! The most fascinating aspects of evolution have never been explained to most of us.

This was barely the tip of the iceberg! I began to uncover an entire body of scientific literature that espouses "non-Darwinian" evolutionary systems. Consider what Neo-Darwinism is, and what it is not:

- Darwin said, "If it could be demonstrated that any complex organ existed which could not possibly have been formed by numerous, successive slight modifications, my theory would absolutely break down." Any theory of evolution that is not essentially gradual is not Darwinism (108).
- Any theory of evolution that incorporates Lamarckian ideas (107), high-speed quantum leaps (113), or any kind of purposeful adaptation (129) is not Neo-Darwinism (643, 645).

I'm not taking a swing at the venerable Charles Darwin. I respect the limitations of what he was able to work with. I admire his foresight. But as we explore Transposition and other evolutionary mechanisms, you'll probably start to agree that the Modern Synthesis of the 1940s was a step backward, not forward.

Equating "evolution" with "Darwinism" also cheats the many great people like Barbara McClintock whose discoveries have moved us away from the Modern Synthesis. McClintock's discoveries suggested a formula more like this:

TRANSPOSITION + NATURAL SELECTION + TIME = EVOLUTION

If you major in biology, you will eventually study Transposition. But it may take some time before you hear about it. Waiting until sophomore college biology to talk about Transposition seems almost like waiting until the second year of medical school to tell obstetrics students to wash their hands.

I found there are two versions of evolution: There's real-world evolution practiced by scientists and medical researchers. And there's a largely fake, dumbed-down version that they bicker about in bookstores and Kansas school board meetings. Had I not sought out sparring partners with my Cosmic Fingerprints emails and resolved to get to the bottom of this, I might have never discovered there was an alternative to what's currently being taught in schools.

I first stumbled upon this because a friend of a friend in 2005 forwarded a link to one of James Shapiro's papers, in which he described McClintock's work, and the amazing protozoan that slices its DNA into 100,000 pieces and rearranges its genome. This struck me as the science story of the century...but for some reason nobody was talking about it. Why? I couldn't believe that with all the books and websites I'd seen arguing about Darwin and evolution and Intelligent Design during the previous year, nobody had ever bothered to mention it.*

Transposition, it turned out, was just the teeny, tiny tip of a huge iceberg. Transposition may be the most common system of evolutionary development, but it's not the only one. And every single one of these mechanisms is modular, contextual, and follows formulaic patterns.

* I also wondered: Could a purposeful kind of evolution also open up an opportunity to drop some of evolution's other 19th-century baggage? Maybe someday evolution might be freed of its "Darwinian" connotations of racism, eugenics, nihilism, and genocide. Had old-school Darwinism become, for some people, their life meta-narrative, the lens through which they see everything that happens? As Richard Dawkins put it: "The universe we observe has precisely the properties we should expect if there is, at bottom, no design, no purpose, no evil and no good, nothing but blind, pitiless indifference. As that unhappy poet A. E. Housman put it: 'For Nature, heartless, witless Nature/Will neither care nor know'" (112). For many atheists, maturity was defined in terms of one's willingness to swallow hard and accept this gloomy reality: "Suck it up and deal with it." Twenty-first-century science paints a far more positive picture of evolution than the gloomy dystopia of Richard Dawkins.

How can cells possibly know what an animal needs and orchestrate a process to make it happen? In chapter 19, you'll see the language and signaling systems cells use to talk to each other. Meanwhile, the next chapter reveals the second blade of Evolution 2.0's Swiss Army Knife.

BULLET POINT SUMMARY:

- Neo-Darwinism says Random Mutation + Natural Selection + Time = Evolution.
- Random Mutation is noise. Noise destroys.
- Cells rearrange DNA according to precise rules (Transposition).

CHAPTER 12

Blade #2: Horizontal Gene Transfer—The Generous Gene

My heart's in a pickle, it's constantly fickle
and not too partic'lar I fear,
When I'm not near the girl I love,
I love the girl I'm near

—Frank Sinatra

ON AUGUST 9, 2012, National Public Radio issued a sobering report about gonorrhea:

> Federal health officials announced that the sexually transmitted infection is getting dangerously close to being untreatable.
>
> As a result, the federal Centers for Disease Control and Prevention issued new guidelines for how doctors should treat gonorrhea. The guidelines are designed to keep one of the remaining effective antibiotics useful for as long as possible by restricting the use of the other drug that works against the disease.

Johns Hopkins researcher Jonathan Zenilman explained that ever since penicillin, we've been caught up in an arms race with bacteria:

> But one by one, each of those antibiotics—and almost every new one that has come along since—eventually stopped working. One reason is that the bacterium that causes gonorrhea can mutate quickly to defend itself, Zenilman said.
>
> "If this was a person, this person would be incredibly creative," he said. "The bug has an incredible ability to adapt and just develop new mechanisms of resisting the impact of these drugs."
>
> Dr. Gail Bolan, head of the Centers for Disease Control STD prevention division, lamented, "We're basically down to one drug, you know, as the most effective treatment for gonorrhea." (135)

How Do Germs Fight Antibiotics?

You're sick and taking antibiotics. The antibiotics cause a bacterium in your body to be immersed in poison. Toxins are penetrating the bacterium's cell walls. It needs to get rid of the poison or it will die.

It was on a short-cut through the hospital kitchens that Albert was first approached by a member of the Antibiotic Resistance.

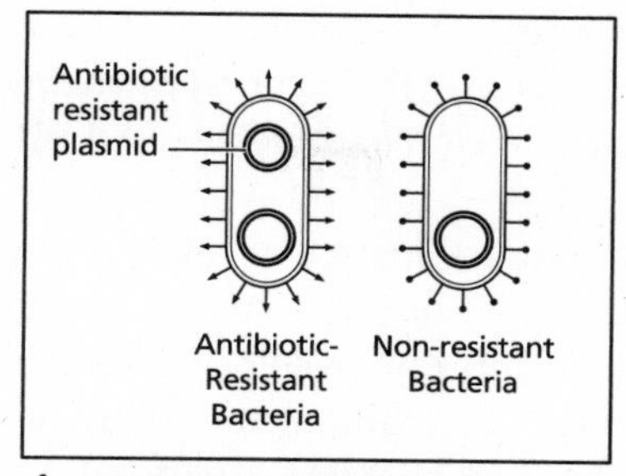

1. An antibiotic-resistant bacterium approaches a normal bacteria.

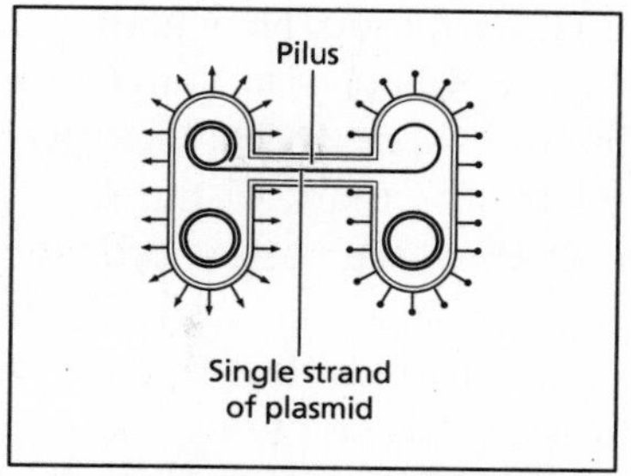

2. The antibiotic-resistant bacterium sends out a pilus to the neighboring bacteria. A single strand of the antibiotic- resistant plasmid is transferred to the normal bacterium via the pilus.

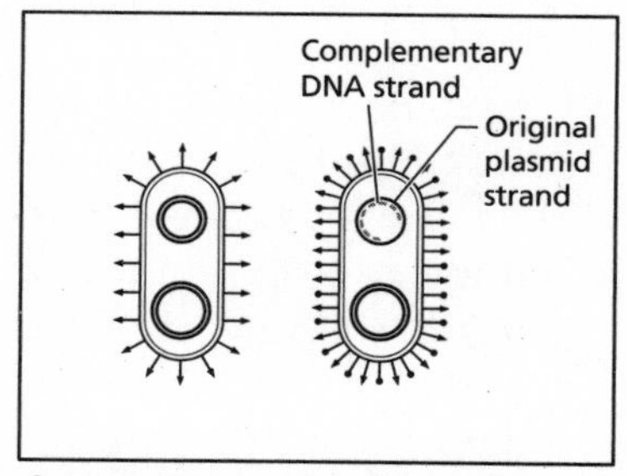

3. After the plasmid strand is established in the recipient cell, a complementary strand is synthesized.

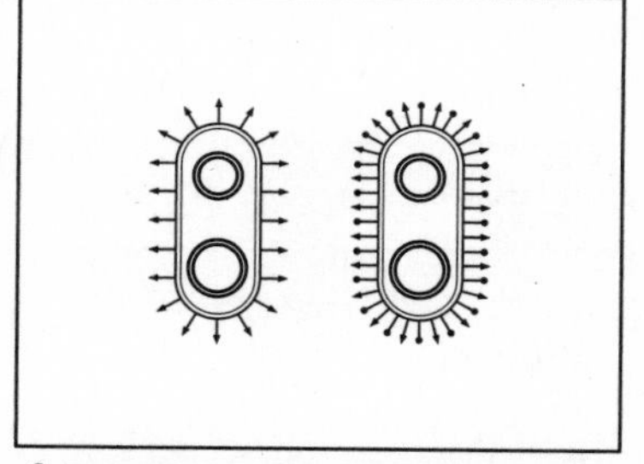

4. Both cells are now antibiotic-resistant.

The bacterium becomes receptive to useful organisms nearby. Those organisms could even be other types of cells, alive or dead. One comes along with a pump that can purge the poison from its own system (255, 615, 228).

When it finds that organism, it receives the organism's plasmid.* A *plasmid* is like a file folder for publicly sharing DNA. The bacterium finds the portion of the DNA that codes for a pump, inserts the new code into its own DNA, and starts multiplying.

Now its offspring sport a pump, too, and the new pump makes the bacteria immune to the antibiotic. All its descendants have immunity

* Pili, which are the germ's tentacles, are equipped with plasmid DNA. They attach to receptors on the recipient, then retract to bring the two cells into contact for DNA transfer. In certain Gram-positive bacteria (types that absorb a crystal violet stain used by biologists), the donors emit a pheromone to elicit stickiness on the recipients.

from that antibiotic. New genetic material spreads through the existing population of bacteria at amazing speed—much faster than the antibiotic can kill them (647). The bacteria that don't adapt, die. All this can happen in less than 30 minutes! This process, known as Horizontal Gene Transfer, is a high-speed adaptive mutation.

Anyone who understands this will be justifiably nervous. You should take comfort in the fact that your own immune system fights the bacteria in precisely the same way, by continually obtaining pieces of DNA that improve its ability to defeat the mutated bacteria. It's literally a contest of cellular military intelligence and evolving weaponry.

The old saying, "Just when you thought you were winning the rat race, along came faster rats," is true all the way down to the cellular level. Welcome to life in an evolutionary world! Nobody gets to be lazy for long.

ONLINE SUPPLEMENT

The war your body wages when you get the flu

www.cosmicfingerprints.com/supplement

Bacteria borrow segments of DNA from other organisms, the same way musicians borrow riffs and melodies from other artists. Just like software engineers borrow blocks of code from other coders. Almost unbelievable! But that's what Horizontal Gene Transfer, the second blade of the Evolution 2.0 Swiss Army Knife, is.

The Nobel Prize–winning biologist Carl Woese (681) was a "celebrity biologist" at the University of Illinois at Urbana–Champaign. A good deal of his fame came from his discovery of a "third kingdom" of bacteria* called archaea. Intrinsic to this was his radical proposal that Horizontal Gene Transfer was the dominant form of evolution when life only consisted of single-cell organisms—meaning that, rather than arising from a single line of accidental mutations from direct ancestors, your DNA is a well-orchestrated mashup of the best features that trillions of trillions of organisms could engineer together over time.

Think back to my conversation with Bob's daughter Melanie: Your immune system is constantly searching for code it needs to fight off invaders. It's constantly discarding old code that it doesn't need any more. Your immune cells make highly accurate predictions of what

* This new "kingdom" was later classified as a new domain.

DNA and antibody combinations will do the job, and then generate them (661). And every one of your cells possesses this same directional ability to edit its DNA. After Horizontal Transfer, Transposition further refines the result.

This is why living things are so finely tuned and optimized for their environments that they often look like they were just "put there" by a divine being!

Cells Distribute Vast *Combinations* of DNA to Other Cells, Too

One bacterium, *Mycobacterium smegmatis*, is related to tuberculosis but is harmless to humans. It doesn't just donate the same DNA to its brothers and sisters. According to an article in *New Scientist*, it uses Horizontal Gene Transfer to endow each recipient cell with a *different* combination of DNA. "We can generate a million [hybrid bacteria] overnight, and each of those million will be different than each other," said researcher Todd Gray (682).

Horizontal Transfer has mostly been recognized for exchanging only small bits of code. According to the article, up to 25 percent of each recipient's genomes were brand new, donated DNA. Before this discovery, scientists had believed such rapid genetic diversity was only possible through sexual reproduction. Scientist Keith Derbyshire said, "I think it's really going to open some eyes about how quickly things can change." (682)

It's impressive that each of the new cells, each having different DNA, are even functional at all. If they were computer programs, you could never just toss big chunks of code into one program from another and get an acceptable result every single time, much less a *better* program. Without precise planning, programs subjected to major edits would just crash.

Yet these bacteria are generating unique permutations of code. They are donating them to sister cells. These sister cells produce new cells that are superior to their parents. Whatever these bacteria are doing is a highly coordinated function of modular programs. This is worthy of much deeper study.

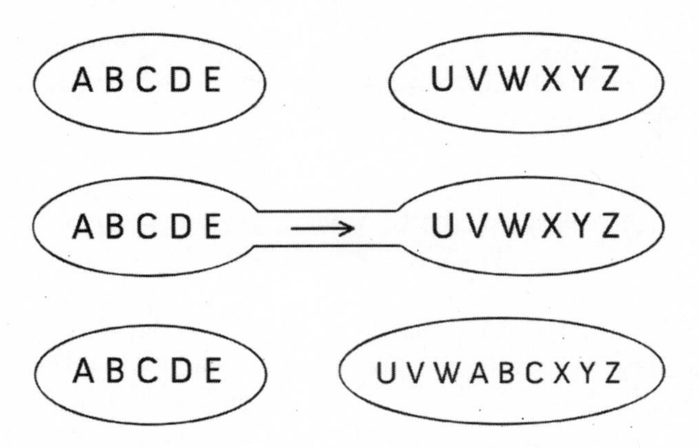

In Horizontal Gene Transfer, cells exchange DNA. A bacterium can "download" and execute new DNA instructions from another cell in as little as 20 minutes. (The adaptation would never happen at all if it depended on Darwinian random mutations.) This is why you have to finish your antibiotics—if you don't kill the infection, it comes roaring back with more ferocity.

Horizontal Transfer and the Tree of Life

Horizontal Transfer means that an organism's genes don't necessarily even come from its parents. They can actually come from... perhaps anywhere! A chunk of code might get passed from a bacterium to a chinchilla to a virus to a fern plant, to a bacterium and back into another animal, because a channel exists for passing genetic material directly between living organisms.

Cells combine newly donated DNA with the DNA they already have. Here's an English analogy of how Horizontal Transfer works, using sentences from childhood:

We combine

The quick brown fox jumps over the lazy dog

with

Horizontal Transfer fundamentally alters our idea of the "evolutionary tree of life." Thin threads connect what we previously saw as unconnected branches, creating a vast web. The connections don't just crisscross—they form loops and dynamic feedback systems.

> They sailed away, for a year and a day, to the land where the Bong-tree grows.

By recombining "genes"—while obeying the rules of grammar—we create all kinds of interesting sentences:

> The quick brown fox jumps over the Bong-tree.
> The quick brown fox jumps over the lazy dog, for a year and a day.
> The quick brown fox sailed away from the lazy dog.
> They sailed away to the land where the quick brown fox jumps over the lazy dog.

The quick brown fox jumps over the land where the Bong-tree grows.

Again, notice how utterly *opposite* from randomness this is. English has spelling and grammar rules, which are necessary for a new construction to make sense. DNA is no different. Like the English language, DNA operates under defined rules to create predictable combinations of amino acids. The mutations have to obey the rules of the language. Otherwise you get legs growing out of your head.

If you begin making substitutions, while making sure you follow the rules of English, you get all kinds of interesting, functional combinations. You create things that didn't exist before. Language development gives natural selection something to work with.

One researcher explains, "Genes are never transferred alone. They are transferred in unit-constructs, known as 'expression cassettes.' Each has to be accompanied by a special piece of genetic material, the promoter, which signals the cell to switch the code on, i.e., to transcribe the DNA code sequence into RNA. At the end of the gene there has to be another signal, a terminator, to end the transcription and to mark the RNA, so it can be further processed and translated into protein" (621).*

In other words, DNA cassettes for transferring code from one organism to another are structured like this:

ATCAAGTCC	CGGAATCAG	TCGTAGCAAT
Promoter	Gene	Terminator

In Ethernet and computer files, data is structured in a comparable way—there is a beginning of a message, the header ("Hello, I've got something to say"); the message itself; and then the end of the message, the footer ("I'm all done now, goodbye."):

001101010011	010111000101	000110101100
Header	Data	Footer

* This is not always the case. In integrons, the individual cassettes have only the coding sequence, and the integron itself has the promoter (674).

You see here that evolutionary steps are not random, mysterious events that somehow take place over millions of years. They are discrete and measurable. And they behave remarkably like our own engineered systems.

Organisms are equipped to share information with each other the same way your favorite coffee shop is fitted with a wireless network and your laptop computer sports a Wi-Fi transceiver. They all speak the same language so they can share information. Horizontal Transfer is known to occur from bacteria to bacteria, bacteria to plant, plant to plant, animal cell to virus, mammal to mammal, and bacteria to invertebrate. No hard limits are known to exist in the transfer of DNA between various categories of organisms (659).

Most "Genetic Engineering" Is Cell Engineering with Humans Helping

When people talk about genetic engineering, most of us imagine biologists splicing genes and reinserting them the way audio engineers used to splice analog reel-to-reel tapes. But that's not how it's done.

In reality, most genetic engineering works by facilitating some form of Transposition or Horizontal Transfer. The organisms themselves do the heavy lifting!

In the world of bacteria, bugs become superbugs in minutes. Just as poets, programmers, and architects borrow ideas from each other, cells exchange code as they search for the tools they need to do their jobs.

Once they have this code, they often keep it even if they're not ready to use it yet, passing it down to future generations.

If it seems outlandish to suggest a "simple cell" can do such a thing, the next chapter paints an even more astonishing picture.

BULLET POINT SUMMARY:

- Neo-Darwinism says Random Mutation + Natural Selection + Time = Evolution.
- Random Mutation is noise. Noise destroys.
- Cells rearrange DNA according to precise rules (Transposition).
- Cells exchange DNA with other cells (Horizontal Gene Transfer).

CHAPTER 13

How Smart Is a Cell, Really?

When the ebbing tide retreats
Along the rocky shoreline
It leaves a trail of tidal pools
In a short-lived galaxy
Each microcosmic planet
A complete society

—RUSH

BEFORE WE DISCUSS how cells keep the code they pick up from other cells, there's something else we need to briefly explore: How can tiny cells know how to rewrite code? If you can't mash up chunks of computer code and get new, useful computer programs, then how do cells know how to reengineer themselves?

So far I've shown you two of Evolution 2.0's five Swiss Army Knife blades: Transposition and Horizontal Gene Transfer. I'm about to show you three more. But before I go on, we need to pause and consider just how smart cells—the actors that possess these blades—really are.

Headphone Cords Tangle; DNA Strands Don't. Here's Why.

One cell can hold a gigabyte of data; plant and animal tissues have a billion cells per cubic centimeter. One juicy bite of steak (5 cubic centimeters) contains over 10^{18} bytes of data, an exabyte—more than all the videos downloaded on the entire internet in a single day.

If you stretched a strand of human DNA end to end, it would be 6 feet (1.8 meters) long. But it is folded into a space so small that it is literally a trillion times denser than any hard drive. Plus the cell easily reads it whenever necessary. How is this possible?

The folding of DNA inside the nucleus of the cell is fractal. A unique mathematical pattern places folds within folds within folds so that, unlike your headphone cord, DNA strands don't tangle (678).

Unused DNA is kept in a high-density area, using the folding pattern within a pattern. This storage pattern is called a *fractal globule*. It enables the cell to store DNA in amazingly little space, avoiding tangles and knots that would destroy the cell's capacity to read its own instructions. The DNA quickly unpacks and repacks during gene activation and cell replication.

"Nature's devised a stunningly elegant solution to storing information—a super-dense, knot-free structure," says senior author Eric Lander, director of the Broad Institute, who is also professor of biology at MIT and professor of systems biology at Harvard Medical School (606).

The globule is a lattice, a pattern known to mathematicians, in which every point is only visited once and no paths intersect. This prevents knots from forming.

The *Harvard Gazette* reports, "The human genome is organized into two separate compartments, keeping active genes separate and accessible while sequestering unused DNA in a denser storage compartment... the information density in the nucleus is trillions of times higher than on a computer chip" (606). Cells move chromosomes back and forth between the two compartments as needed.

Steve Jobs and the Calligraphy of Life

The late CEO of Apple, Steve Jobs, was asked to speak at a Stanford graduation ceremony in 2005. He told the story of wandering into a calligraphy class in college—a totally serendipitous experience. He fell in love with calligraphy and because of this, he equipped his first computer, the Macintosh, with graphic capability and beautiful fonts.

He wanted every student there to understand that sometimes it's the unexpected paths that contribute the most to your life.

If the genetic code is text, proteins are calligraphy. All cells and tissues are built from proteins. They're like beautiful fonts that transform mere assembly instructions into art. In the same way that a Mac in the '80s was head and shoulders above all its drab cousins in the computer world, proteins are not just building blocks; they, too, are works of art.

The structure of those proteins comes from yet another layer of coding in the digital message. I purchased a copy of *The Protein Chart,* which is like a periodic table of these beautiful proteins. If you ever thought of proteins as "blobs of organic compounds," you'll never think of them that way again. Here's a small sampling of these remarkable assemblies:

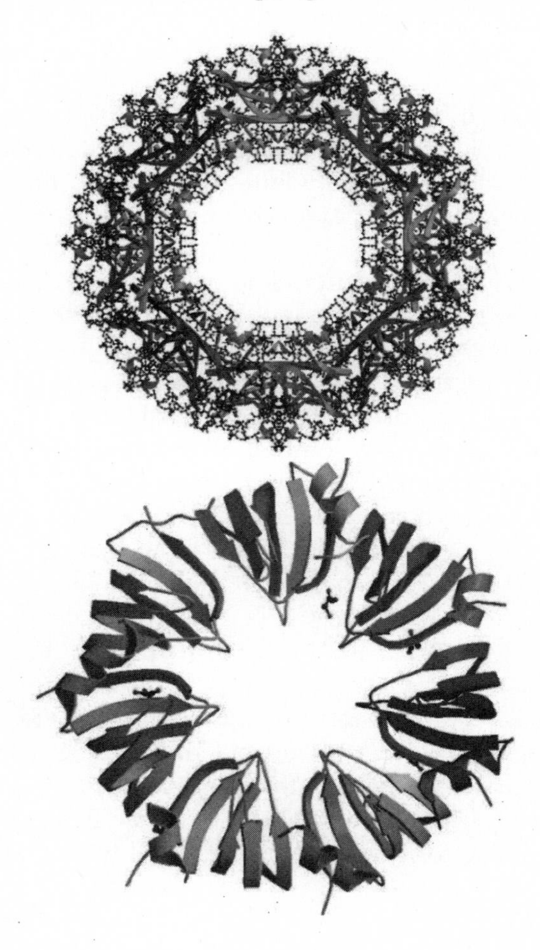

There are about a million proteins; each one is different. They all have fascinating geometrical symmetries like the ones you see here. And that's only the beginning of the cell's wonders.

99.9999999% Accurate

E. coli bacteria duplicates its 4.6-megabyte genome in 40 minutes. It achieves phenomenal 99.9999999% copying precision (nine 9's) in three stages.

The molecular machine that performs the first copying step makes 1 mistake every 100,000 letters. Sensor-based proofreading then adds a second step. An enzyme senses distortion of the DNA helix from incorrect letter insertions and halts polymerization. It removes the incorrect base, inserts the correct letter, and resumes operation. This multiplies accuracy by a factor of 100 to 1,000.

Last, the cell employs three proteins for final proofreading. They do *mismatch repair,* clipping out sections of DNA that contain erroneous letters and inserting newly manufactured, error-free DNA. A special methylation feature prevents the machines from confusing the original DNA with the newly copied DNA. This stage further multiplies copying accuracy by a factor of 100. These three stages of error correction cascade, for an accuracy of one mistake per billion letters. (659)

When I first heard about this, I found it hauntingly similar to the many stages of error correction in computers and networking components. Hardly anybody thinks or talks about them, but those programs quietly do their work 24/7.

Cells Make Decisions as Members of a Superbly Organized Army

The authors of a *Newsweek* article entitled "The Secrets of the Human Cell" explained,

> Each of those 100 trillion cells [in the human body] functions like a walled city. Power plants generate the cell's energy. Factories produce proteins, vital units of chemical commerce. Complex transportation systems guide specific chemicals from point to point within the cell and beyond. Sentries at the barricades control the export and import markets, and monitor the outside world for signs of danger. Disciplined biological armies stand ready to grapple with invaders. A centralized genetic government maintains order. (217)

Cells do their work silently, processing prodigious volumes of information with tremendous speed. Carl Sagan wrote,

> A living cell is a marvel of detailed and complex architecture. Seen through a microscope there is an appearance of almost frantic activity. On a deeper level it is known that molecules are being synthesized at an enormous rate.
>
> Almost any enzyme catalyzes the synthesis of more than 100 other molecules per second. In ten minutes, a sizeable fraction of total mass of a metabolizing bacterial cell has been synthesized. The information content of a simple cell has been estimated as around 10^{12} bits, comparable to about a hundred million pages of the *Encyclopedia Britannica.*" (225)

More Than Just Blobs of Protein

We've been taught to imagine cells as just tiny little blobs of protein that make copies of themselves. At best, perhaps we've thought of them like blind, unthinking computer programs, receiving instructions and acting on them like robots—little more than tiny chemical factories. Slowly, I came to realize that if you had superpowers and you could shrink yourself by a million times, you'd be exhilarated by an entirely different picture.

You may well describe a cell as a tiny, self-intentional supercomputer with dozens of sensors, constantly processing information about food, temperature, toxins, sister cells, predators, sexual partners, pheromones, and courtship rituals. The cell responds to all this information gathering with elaborate communication systems and editing abilities. Yet this amazing description still falls short (659). Organisms store this information in DNA and RNA databases, and in molecules all across the cell.

There's an amazing TED video by Dr. Bonnie Bassler of Princeton University. It's called "How Bacteria Talk" (604) and you can watch it online at www.cosmicfingerprints.com/intelligent-bacteria. In this video, she describes how bacteria generate light, but only when working together in groups:

> What was actually interesting to us was not that the bacteria made light, but when the bacteria made light. What we noticed is when the bacteria were alone, so when they were in dilute suspension, they made no light. But when they grew to a certain cell number all the bacteria turned on light simultaneously.
>
> The question that we had is how can bacteria, these primitive organisms, tell the difference from times when they're alone, and times when they're in a community, and then all do something together. What we've figured out is the way that they do that is they talk to each other, and they talk with a chemical language.*

* One might form the impression from the video that Bassler's team discovered bioluminescence and the bacterial ability to self-transform. These discoveries were made some 35 years ago—this information is not new! Scientists have known this for decades. Rather, Bassler's contribution in this TED talk is a fascinating, if very abbreviated, description of how bacteria communicate within their colonies.

Seeing Dr. Bassler's video was like looking through a keyhole and finding a lost planet. If bacteria were tiny little "selves" with the capacity to speak to each other using language—if, as we've seen, bacteria edit their own code using complex linguistic rules—then... what else didn't I know about cells?*

The famous biologist Ernst Haeckel described cells in the following way, as reported by a writer in *Nature* in 1873: "small formless masses of albuminous combinations of carbon, and differing from each other only in their mode of reproduction, development and nutrition. As these living beings do not present any complication of diverse parts, any division of functions or of organs, as all the phenomena of life with them proceed in a homogeneous manner, and without determinate form, it is very easy to conceive of their spontaneous generation" (226).

Though he was proven wrong a century ago, his flippant attitude lingers even now. Two of the largest mistakes in 20th-century science are (1) random mutations in evolution, always assumed but never proved (and mathematically impossible to prove); and (2) greatly underestimating the capabilities of the basic unit of life: the cell. It purposefully engineers adaptations to achieve its own goals (645).

* In the field there are two journals devoted to biosemiotics (semiotics is the study of signs and language). Numerous linguists have weighed in on major problems in immunization and genomics (515). We know that instructions to build entire organelles are exchanged between cells in Horizontal Gene Transfer. Chemical communication between cells is considered by many biologists to be linguistic (405, 520). Currently we understand the syntax ("grammar") of cellular communication reasonably well, but our analysis of semantics ("meaning") is much less clear.

Those who are skeptical of these claims should begin with Bassler's TED video. Check the references cited here and search http://scholar.google.com/ for *DNA* with combinations of the following words: *genome*, *linguistic*, *linguistics*, *semiotic*, *syntax*, *semantics*, *pragmatics*, and *universal grammar*. While the semiotic (linguistic) school of thought is presently a minority view in biology, one can easily verify that it is supported by substantial published research (520).

If we consider that a strand of DNA has segments that (1) symbolically code for proteins, (2) contain symbolic instructions for assembling those proteins in three-dimensional space, (3) contain instructions that dictate the timing of those events, and (4) feature epigenetic systems that switch sections of code on and off to build different tissue types like liver, skin, or heart cells, we have a case for a four-layer communication system—four levels of abstraction, or stacks of Russian dolls. Even that description is grossly oversimplified; cells have the ability to edit all these instructions and adapt to new situations. One need not settle for taking a linguist's or biologist's word for it; the most essential features of the DNA language are plainly observable in the basic functions of the genetic code.

Since DNA is demonstrably linguistic (520), we have good reason to entertain the hypothesis that other forms of cellular communication are linguistic as well.

I suspect that cellular language may prove to be *more* sophisticated than human language. I predict that we stand to learn more by studying the cell than any other topic in science. It's time to stop selling the cell short.

Bacteria Use Sophisticated Language

Bacteria are the most abundant kind of cells. Most things bacteria do, other kinds of cells do, too. Our knowledge of bacteria forms the foundation for understanding all forms of life.

Across the vast kingdoms of life on Earth, bacteria's linguistic skills are second only to humans'—and in some ways they're superior. In her TED talk, Dr. Bassler goes on to describe how bacteria send out very small molecules, which form "words" for communication. Each molecule in the chain acts as a letter in the word. Different words form commands or requests, which are understood by their neighbors. Each molecule in the chain acts as a letter in the word (400, 604).

Günther Witzany, a German philosopher of biology (407), also explains that bacteria interpret signals based not just on the chemistry but on the context of the situation. Bonnie Bassler also described bacterial communication as linguistic (401).

Moreover, bacteria have molecular words for "me," "you," "us," and "them." They know and describe the difference between themselves and others. They sense how many of their own species and how many of another exist in any population. And they speak multiple languages—a native language for their own species, and foreign languages for other species.

To conduct intraspecies conversations, a bacterium has a special receptor that only accepts molecules from the bacterium they're conversing with. Dr. Bassler reports that "each molecule fits into its partner receptor and no other. So these are private, secret conversations" (604).*

* Anthropomorphism—describing cells in ways that suggest humanlike qualities—has traditionally been frowned upon in biology. Yet ask yourself: Which of the following things is a cell most like: (a) rock (b) sand dune (c) star (d) lake (e) molecule (f) human (g) computer (h) combustion engine? Since rocks, snowflakes, crystals, planets, and other nonliving things do not self-replicate, do not contain codes, do not repair themselves, and do not seek nourishment, the known physics of nonliving things is inadequate for the task of describing life. Prohibitions against such language are motivated by the desire to purge teleological ideas from science (525), and I see no good reason to forbid such comparisons. Other authors argue in favor of anthropomorphic language as well (508).

But alongside this, bacteria have a second, generic receptor that emits a different molecule and allows them to communicate with members of different species of bacteria. Dr. Bassler describes this as "bacterial Esperanto," a language constructed for universal communication. She says bacteria will make decisions based on which species is in the majority and minority of any given population (604). As the paper "Bacterial Linguistic Communication and Social Intelligence" (401) points out, "Bacterial chemical conversations also include assignment of contextual meaning to words and sentences (semantic) and conduction of dialogue (pragmatic)—the fundamental aspects of linguistic communication."

Bacteria Work in Teams

This capacity for communication makes bacteria social creatures (513), and, as it turns out, democratic ones as well. Dr. Bassler says, "They make chemical words, they recognize those words, and they turn on group behaviors that are only successful when all of the cells participate in unison. We have a fancy name for this; we call it 'Quorum Sensing.' They vote with these chemical votes, the vote gets counted, and then everybody responds to the vote" (604). Apparently, the practice of casting votes wasn't invented by Benjamin Franklin or the Greeks. It was invented by bacteria!

I used to think of bacteria as "lone rangers." But seldom do bacteria float around all by themselves groping for something to eat. They live together in colonies with assigned roles and allegiance to the group. They behave much like ants or bees (502, 513). They greet each other when they meet, they hunt for food together, and they pool their digestive enzymes (522). They prey on other bacterial colonies by surrounding them, digesting them, and splitting the booty (much as human armies do).

Martin Dworkin, a preeminent scholar in microbiology at the University of Minnesota (524), said, "In the presence of clumps of prey bacteria, swarms of *M. xanthus* would frequently turn sharply, head directly for that clump, and then linger there as if at a banquet." He called it "the microbial wolf-pack effect." He also showed how bacteria can detect and migrate toward glass beads, but won't stay and eat as they do after locating a clump of prey bacteria (504).

Certain kinds of cyanobacteria called *Anabaena* divide labor by forming cells called heterocysts. When deprived of nitrogen, they genetically reengineer their genomes to form an enzyme that extracts single nitrogen atoms from the air (507). In other words, division of labor wasn't invented by Henry Ford or Adam Smith or ancient tribesmen—it was invented by bacteria.

When bacteria want to attack you and make you sick, they don't just hack their way into your system and try to take over. Any one bacterium is far too small and helpless to pull this off. Rather, they wait until they have built up to a sizable population. They begin signaling each other and when they estimate they've reached a critical mass, they launch an organized attack (516).

A paper in *Nature* (514) by Iñigo Martincorena, a fellow at the Sanger Institute, and colleagues reports that mutation rates vary widely from one place in the genome to another. Factors we don't yet understand influence mutation rate of cells, and mutations maximize the chances of survival. These findings contradict the current popular belief that mutations occur regardless of their ability to help the organism survive.

So cells direct their own evolution. They also monitor their genomes to find and fix mistakes. Which brings us to the next surprise.

Long-Term Planning and Self-Sacrifice

In plants and animals, cells obediently die in response to instructions from other cells. Programmed cell death protects against cancer (see chapter 26). When embryos form, cells between fingers die so each finger can move independently (503). Plant cells infected with bacteria or viruses die on command to prevent disease from spreading. In bacterial colonies, single cells die to maintain the genetic stability and function of the group. They do this in response to chemical signals (505). Dead cells are disassembled and replaced with new cells.

Cells are able to make other changes that are disadvantageous in the present but expected to be helpful in the future based on recognition of past patterns. They weigh the costs and benefits of future decisions (521, 236). They make sacrifices for the common good and respond differently to various threats.

The cells that make up plants and animals are just as intelligent as bacteria, but they rigidly follow the agenda of the entire organism. We have a name for cells that break rank and develop their own separate identity: cancer. I talk more about this in chapter 26.

Are Cells Self-Aware?

Barbara McClintock, who discovered Transposition in her corn experiments, was the first biologist to ask, "What does a cell know about itself?" (608).

Are cells sentient? I truly do not know the answer. The idea that they could be self-aware makes most of us squirm. Perhaps it's not a simple "yes" or "no" answer. Perhaps living things have various degrees of self-awareness. But if there's any truth to the idea that "nature is as nature does," then you can't help but wonder.

Evolution is ultimately driven by cells' desire to multiply, to fill the Earth, to use every available resource to its maximum potential, and to populate every ecological niche with fantastic beauty and diversity.

Biology Is as Biology Does

Someone will inevitably object: "We have no reason to believe cells are intelligent. They are too small, they don't have brains, and we don't know how this could even be possible."

In science it is never necessary to explain *why* something is true. We only need to observe that it *is* true. We don't know exactly why gravity is true, though we'd sure like to, but our ignorance about its inner workings doesn't make it false.

Albert Szent-Györgyi, winner of two Nobel Prizes, explained why complex interactive systems don't evolve by accident. He insisted individual parts must simultaneously coevolve: "Saying it can be improved by random mutation of one link, is like saying you could improve a Swiss watch by dropping it and thus bending one of its wheels or axes. To get a better watch, all the wheels must be changed simultaneously to make a good fit again" (250). Every engineer knows this.

Biology is as biology does. If cells act intelligently, then why not accept that they're intelligent? Let's take the evidence at face value. If it takes the next 100 years to discover how this could be so, then the century ahead of us promises to be enthralling. To accept this view requires us to embrace a higher understanding of nature. We *always* tend to underestimate nature. Nature will *always* surprise us. Nature *always* has another treasure waiting for us.

Replace randomness with the highly coordinated, goal-seeking mechanisms we're exploring here. Twenty-first-century research shows that cells engineer their own evolution. Suddenly the whole adaptive framework sparkles with new color. *The watch has the capacity to redesign itself.*

BULLET POINT SUMMARY:

- Neo-Darwinism says Random Mutation + Natural Selection + Time = Evolution.
- Random Mutation is noise. Noise destroys.
- Cells rearrange DNA according to precise rules (Transposition).
- Cells exchange DNA with other cells (Horizontal Gene Transfer).
- Cells communicate with each other and edit their own genomes with incredibly sophisticated language.

CHAPTER 14

Blade #3: Epigenetics—How Parents Pass Learned Traits to Their Kids

I said, "I'd like to see you if you don't mind."
He said, "I'd love to, Dad, if I could find the time.
You see, my new job's a hassle, and the kids have the flu,
But it's sure nice talking to you, Dad.
It's been sure nice talking to you."
And as I hung up the phone it occurred to me,
He'd grown up just like me.
My boy was just like me.

—HARRY CHAPIN

IN THE DUTCH FAMINE OF 1944 during World War II, thousands of unborn children experienced harsh deprivation, which resulted in unexpected changes. Not only were the children who were in utero during the famine smaller; when these children grew up and had children, their children were also smaller than average. Over their life-time, the children of the famine experienced far-above-average rates of obesity, type 2 diabetes, cardiovascular problems, and other diseases related to an unhealthy body weight.

This data suggests that the famine experienced by the mothers triggered *epigenetic* changes (220) that were inherited by the next generation—in this case, slower metabolisms and smaller statures that would allow them to better survive conditions where there was a lack of food. But those same adaptations led to metabolic diseases in a post-famine environment where food was readily available.

Epigenetics doesn't just alter the metabolism of children conceived in times of famine, though. It controls the expression of every cell in your body. The reason you have hair on the top of your head and not your forehead is because epigenetic factors control the expression of hair differently in different areas of your skin. It's also the reason identical twins with identical DNA can still have different allergies, intelligence, aptitudes, and even different inherited diseases.

Epigenetics is blade #3 of the Evolution 2.0 Swiss Army Knife. It's a switch that "grays out" genes, altering DNA's function without changing the DNA sequence itself. It produces different cell types in fetal development; it alters tissues based on the external environment, and passes learned traits to offspring. Coding sequences stay the same but their expression is altered through a combination of mechanisms (248).

Only 10 to 20 percent of genes are switched on in any cell (622). For example, the many genes for eye color express themselves in your eyes, not necessarily in your liver or tongue. Epigenetic processes are not necessarily fixed over the lifetime of an animal or person, either. Epigenetically controlled factors in cells change over time, over the course of your life, in response to the environment. The genes of elephants living in erratic periods of drought may switch on and off to conserve water. The genes of an athlete may change states to build smoother or striated muscle mass, depending on the kind of exercise they're doing. You have a fixed number of muscle cells; what increases when you exercise is the amount of specific proteins in your muscles, and that is controlled by Epigenetics.

In fact, genes in animals and plants switch on and off on a daily basis (651). For example, cells regulate their energy consumption. They store more fat when you have too much food, or use fat as a nutrient when there is not enough food to sustain you. External stimulus alters their response.

When your environment changes, so does your cells' environment. Even when the change is something as minor as too little sugar in your bloodstream, cells respond by changing their code expression. Removing pieces of code altogether would be overkill. They might need that code later. So instead, cells have ways of temporarily shutting down activation in response to external situations.

A recent project that documented epigenetic change in real time was done on rat pups. When rat mothers give birth to their pups, they clean and groom them. Researchers in Canada discovered that pups licked by their mothers had a reduction in a specific type of gene expression in the hypothalamus. This in turn provoked a chain reaction, which caused the pups to handle stress more successfully. In other words, the physical action of their mothers altered code expression in a specific organ, without changing its DNA (683, 645).

Plain-English Example of Epigenetics

I can take a sentence and make it say the opposite of what it said before, just by selectively "graying out" words and phrases:

> Flight 6429 was delayed from Chicago to Winnipeg so it will not be departing before 6:45 P.M.
>
> Flight 6429 was delayed from Chicago to Winnipeg so it will not be departing before 6:45 P.M.
>
> ("Flight 6429 was from Winnipeg.")
>
> Flight 6429 was delayed from Chicago to Winnipeg so it will not be departing before 6:45 P.M.
>
> ("Flight 29 to Winnipeg will depart before 6.")

This illustrates how Epigenetics works—by "graying out" code sequences and making them silent. The mechanism that grays out the code is called *methylation*. Methylation is how different cell types get built from the exact same strand of DNA. One methylation pattern activates the genes for building neurons; another pattern builds muscle tissue; another builds skin cells.

Methylation is an ingenious form of data compression, because multiple epigenetic templates can generate completely different messages from the same sentence. Not only that, the organism can shuffle around grayed-out code to work out further adaptations. During this time the silent sections of DNA are like random access memory in a computer—extra space where secondary jobs get done without interfering with the current business of the organism.

And of course, in order for this to work, the epigenetic mechanism must observe the grammatical rules of the DNA language.

The Evolutionary Picture Comes into Focus

We've now established that the cells in corn plants and every other organism in existence have the ability to rearrange their own DNA. Bugs become superbugs as bacteria cells exchange DNA with each other and with the cells of other organisms, and they possess built-in, exquisitely sophisticated machinery for doing this. We've also seen that parent cells can gray out code in their genes and their daughter cell's genes to help better adapt to current threats.

You play the guitar and get callused hands. Those calluses are an epigenetic response to a call for thicker skin. That doesn't mean your next baby will have calluses on his hands too. Still, sometimes epigenetic changes are passed to offspring as in the Dutch famine, driving evolution. Which changes get passed on, and exactly how, is the focus of ongoing research.

Why Isn't Anybody Talking About This?

None of this information is just lying around for average folks to pick up. I had to dig pretty deep to find this stuff. Not that it's impossible; it's just that if you didn't know what to look for, you would not likely find it.

The deeper I go, the more I find these discoveries support adaptive models that Darwinists rejected almost a century ago.

What Does a Plant Know?

The cells of a cherry tree decide to bloom on a day in the spring, but they don't bloom again on an identical warm day the next fall. That's because they use Epigenetics to record their memory of winter. They pass the memory of that winter to their offspring (610).

Plants coordinate cell activities with a complex network of electrical, chemical, genetic, and pheromone signals. Each cell is doing a specific job in a specific context, supporting the health of not only the plant but its neighbors. From Daniel Chamovitz's fascinating book *What a Plant Knows: A Field Guide to the Senses*:

> Plants see you. In fact, plants monitor their visible environment all the time. Plants see if you come near them; they know when you stand over them. They even know if you're wearing a blue or a red shirt. They know if you've painted your house or if you've moved their pots from one side of the living room to the other.
>
> Plants perceive light because they have color receptors for blue and red light. Color receptors tell the plant the length of days and seasons. They use this information to judge when to bloom, when leaves should change and when to go dormant for the winter.
>
> When a lima bean plant is eaten by beetles, it responds in two ways. The leaves that are being eaten by the insects release a mixture of volatile chemicals into the air, and the flowers (though not directly attacked by the beetles) produce a nectar that attracts beetle-eating arthropods.
>
> Simply touching an arabidopsis leaf [related to mustard and cabbages] results in a rapid change in the genetic makeup of the plant... An arabidopsis plant that's touched a few times a day in the lab will be much squatter, and flower much later, than one that's left to its own accord. Simply stroking its leaves three times a day completely changes its physical development. (610)

Chamovitz is describing the plant's epigenetic response to its environment. Environmental signals initiate subtle heritable changes to offspring.

Lamarck's Revenge

Sixty years before Darwin published his famous book, French biologist Jean-Baptiste Lamarck proposed that an organism can pass acquired characteristics to its offspring. Darwin himself embraced some of

Lamarck's thinking (108, 645). But after Darwin, Lamarck was mocked and ridiculed and his theories were tossed out. Lamarck's ideas offended scientists who proclaimed that evolution was blind and purposeless.

MIT Technology Review vindicated Lamarck in 2009, reporting on the rat pups experiment: "The effects of an animal's environment during adolescence can be passed down to future offspring... The findings provide support for a 200-year-old theory of evolution that has been largely dismissed: Lamarckism, which states that acquired characteristics can be passed on to offspring" (667).

Could Epigenetics be a reasonable explanation for why bears know how to hibernate, salmon know where to spawn, and birds know how and where to migrate and build nests?

Modern research increasingly suggests that animals can pass on learned traits to their offspring and that there really is a "memory" of past events overlaid onto DNA (667). This memory is flexible and adaptive.

This raises all kinds of questions: If you overeat, do you pass your obesity on to your children? What about alcoholism and family dysfunctions like phobias and anger?

The extent and power of Epigenetics is still a subject of debate. But what the research so far tells us is that Lamarck discovered a key truth. Parents do pass certain acquired traits to their offspring.

From what I've said so far, it might seem as though Epigenetics is primarily a tool of fine-tuning and gradual adaptation over multiple generations, and it is. But in concert with Genome Duplication—blade #5, which I'll describe in chapter 16—it becomes an agent of massive change and high-speed progress.

First, though, let's look at another mechanism of Evolution 2.0, and how it was discovered by Russians in the early 1900s, lost and nearly extinct—then excavated and brought to life in the last 30 years.

BULLET POINT SUMMARY:

- Neo-Darwinism says Random Mutation + Natural Selection + Time = Evolution.
- Random Mutation is noise. Noise destroys.
- Cells rearrange DNA according to precise rules (Transposition).
- Cells exchange DNA with other cells (Horizontal Gene Transfer).
- Cells communicate with each other and edit their own genomes with incredibly sophisticated language.
- Cells switch code on and off for themselves and their progeny (Epigenetics).

CHAPTER 15

Blade #4: Symbiogenesis—Evolution as Cooperation

A little bit of me
And a whole lot of you
Add it up together
And here's what you're gonna do

—311

FROM 1997 TO 2001, I worked for a tech company in Chicago. We sold circuit boards that enabled devices like bar code readers and temperature controllers to communicate over a factory network. Because our product was too costly for many of our customers, we designed a $25 chip that would replace six or seven $200 circuit boards and occupy one-tenth of the space. This chip was a major evolution, a quantum leap. We were able to create it quickly by sourcing the components we needed from other companies rather than trying to reinvent them.

Our new chip came with a surprising bonus: A large, publicly traded company had been trying to develop an almost identical chip, but their team was a year behind us. When our chip was completed, they bought our company.

They acquired everything—not just the chip, but our other product lines, intellectual property and trademarks, sales and distribution channels, and all the employees. During the following year they integrated the operations of the two firms, getting rid of duplicate functions and reassigning employees to new tasks.

Nature initiates the same process of integration. In biology, the process is called *Symbiogenesis*. Yeah, that's a clunky six-syllable word, but its roots are simple: *symbio* = cooperation, *genesis* = creation. Cooperative creation. Different kinds of cells merge to create new cells, or two organisms merge to create a new one—bringing their reproductive cycles and their physiology together in lockstep.

Our company's new chip was the product of Symbiogenesis: We tightly packaged a range of separate components together into a new single component. This is precisely what *integrated circuit* means. The other company acquiring our company and merging all the departments into a new organization was another kind of Symbiogenesis.

Living organisms think the same way as the parent company with hundreds of employees did when it swallowed up our 15-person startup firm. When cells, plants, or animals need something, they look for it and integrate it into their system. They combine two old things to make something new and better.

Conceptually, Symbiogenesis really is simple—so simple a fifth grader can understand it. In all its myriad gory details, however, it's outlandishly complicated. Our engineers spent two years knitting all these circuits together, painstakingly testing them, and making sure they all worked exactly as they should before our chip went into production. This complexity speaks to the amazing, intentional smarts of these tiny cells.

Symbiogenesis can create sudden, quantum leaps. Symbiosis is also one of the most pervasive themes in all of nature. Consider how many plants and insects live in mutual dependence—like bees being interdependent with flowers. This pattern of symbiosis goes far deeper than this; we find it at the cellular level in all plants and animals.

Symbiogenesis is blade #4 of the Evolution 2.0 Swiss Army Knife. Let's look at some examples.

Examples of Symbiogenesis

Lichen

Discovering symbiogenesis is like noticing a GMC engine is remarkably similar to a Chevy engine, and concluding that one was derived from the other. The first person to develop this concept was Swiss botanist Simon Schwendener in 1867, when he hypothesized that algae and fungus literally merged in the past to form a third organism, a brand new species: lichen (654).

Lichen thrives in extreme environments where neither algae nor fungus can exist. So even though it is possible to dissect lichen into separate components, for all practical purposes it functions as a unique species. In advanced Symbiogenesis, a merger is accomplished between algae and fungus that is so thorough the two cannot be separated.

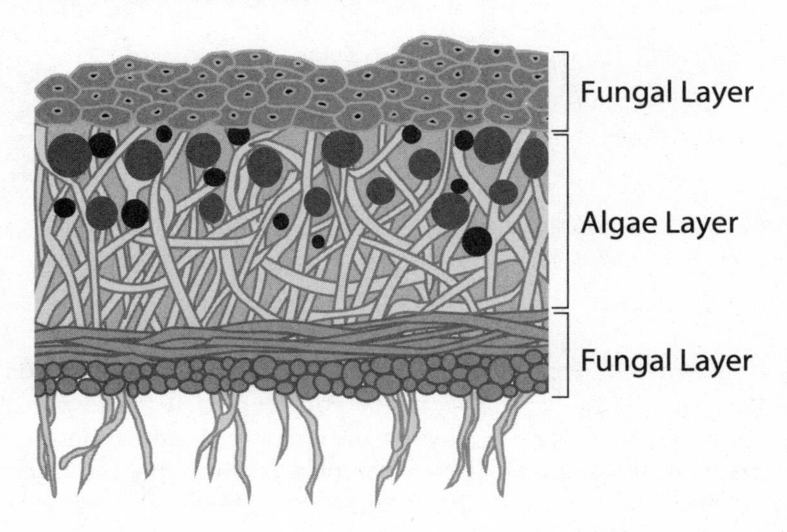

In lichen, algae and fungus don't merely "hitch a ride together." The fungus organizes into a protective layer for the algae, which in turn provides it with nutrients.

Eukaryotes: Complex Cells

In high school biology you probably learned that chloroplasts are the part of the plant cell that conduct photosynthesis. Chloroplasts are separate from the cell itself. They even have their own DNA. Technically, they are cyanobacteria (blue-green algae) living in a symbiotic relationship with a host cell (636). Symbiogenesis theory says chloroplasts originated when a protozoan ingested a cyanobacterium (637). Unlike the lichen in our last example, chloroplasts can't be separated from the host plant cell and survive.

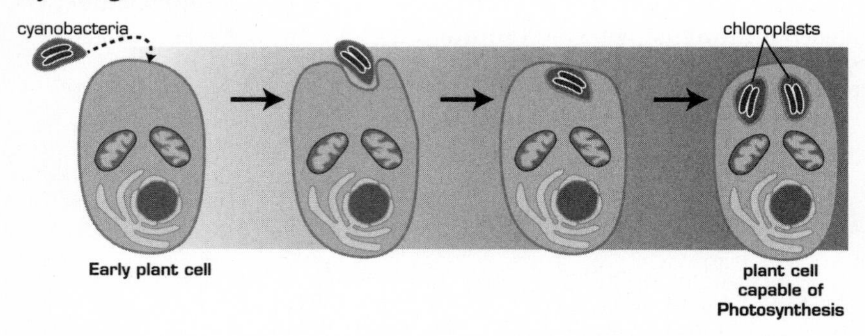

Chloroplasts are algae that carry their own DNA and reproductive machinery. They are embedded inside a larger plant cell. Symbiotic algae give every green plant its color and convert sunlight to energy.

A similar symbiogenesis event made animal cells possible (637). Mitochondria are the organelles inside your cells that are responsible for respiration; they convert oxygen into energy. A mitochondrion is a bacterium that was ingested by a complex cell and, instead of being digested, became its system for respiration. Mitochondria have their own DNA, just as chloroplasts do.

There is nearly universal agreement that both chloroplasts and mitochondria originated through Symbiogenesis. This is firmly established by ribosomal RNA sequencing (652).*

* Most instances of cells ingesting other cells in the lab are complex cells ("eukaryotes") ingesting simpler cells. But simple cells ("prokaryotes") also form symbiotic relationships. In fact, new studies of Symbiogenesis are blurring the lines between prokaryotes and eukaryotes (612).

The Symbiogenesis theory existed 60 years before genome sequencing was common, but the physical similarities between bacteria and mitochondria were already quite apparent. Today we see that DNA sequences of bacteria and mitochondria, and cyanobacteria and chloroplasts, are so strikingly similar that the conclusion is unavoidable. Body parts and code are virtually identical to their symbiotic ancestors, just like circuits in our chip were sourced from their stand-alone cousins.

From now on, every time you see a green leaf or blade of grass, you can thank the blue-green bacterium that merged with a complex cell long ago! Their partnership may be the most successful in all of history.

"Eureka!" Moments, Not Gradual Transitions

Multicellular mergers occurred through a number of stages, but single-celled mergers were consummated in a single step. There's no intermediate form between a protozoan and a blue-green algae becoming a cell that does photosynthesis. Cell mergers overturn the Darwinian doctrine of a thoroughly gradual, continuous transition from one species to another.

Instead, Evolution 2.0, under Symbiogenesis, proceeds through a chain of "Eureka!" moments, when organisms merge successfully. This would explain why the fossil record so often shows a series of sudden jumps. It's also one reason why there aren't nearly as many transitional forms as Darwin predicted.

There is no transitional form between algae and lichen, and there is no half-merger of two cells. Nature loves shortcuts.

Living organisms are just like you and me: They like taking the path of least resistance. Why build it when you can buy it? Why invent it when you can walk across the street and take it off the shelf?

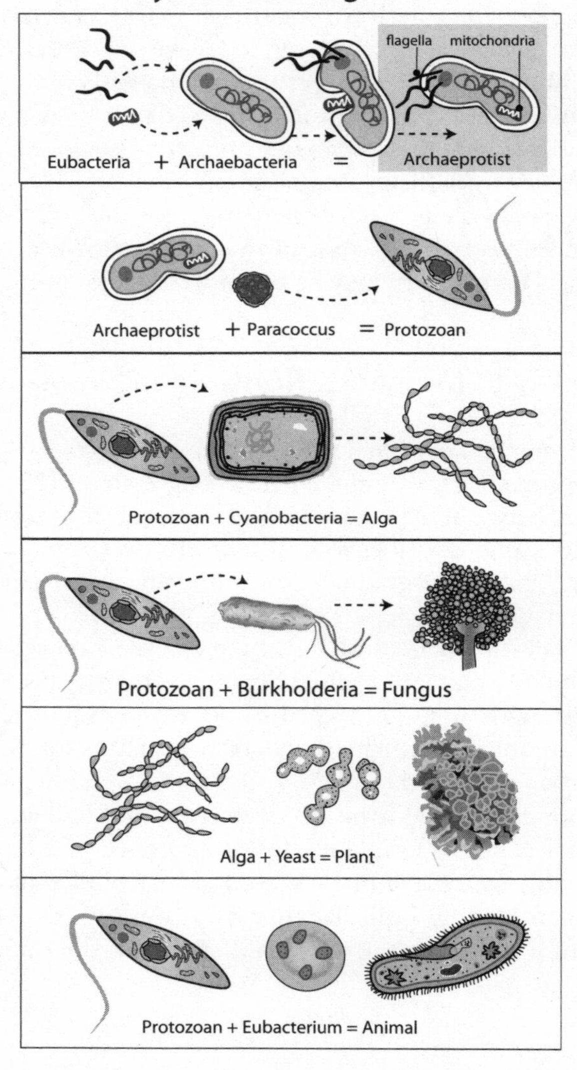

Major classes of cells, plants, and animals are built from symbiotic mergers of multiple smaller organisms. These organisms have DNA and features nearly identical to those of their free-living cousins (after Margulis 637).

SYMBIOGENESIS - HOST/PARASITE

BACTERIA IN YOUR GUT

SYMBIOGENESIS - SYSTEM IN A SYSTEM

In Symbiogenesis, cells merge to form integrated cooperative relationships.

Cells in Symbiosis Seek to Eliminate Redundant Parts, Work More Efficiently

In Symbiogenesis experiments, members negotiate biological functions and exchange portions of DNA, becoming mutually interdependent. Eventually, they adapt such that either can survive on its own.

Watershed events like *protozoan + blue-green algae = plant cell with chloroplast* are believed to have only occurred a half-dozen times in life's history (609). These events completely transformed the organisms involved, endowing them with capabilities they didn't have before. That's why Symbiogenesis events are special and hard to observe. But genomes do record the story of the merger.

It's like finding the same chassis in two cars, the Toyota Camry and the Lexus, with only minor differences in the mounting brackets, and striking similarities down to the most trivial details. One would naturally conclude the chassis was only developed once, then the other car borrowed the design.

It's also like finding the instructions for building the chassis (genetic code) and finding that the instructions, too, are virtually identical. When a schoolteacher gets identical essays from two different students, she never assumes they both came up with two identical paragraphs in a row. She rightly assumes one copied it from the other.

Viruses Are Symbiotic, Too

Most people are familiar with viruses and how they hijack cells, harnessing the cell's machinery to replicate like crazy. But there's another kind of virus, called an *endogenous retrovirus,* that works very differently. It inserts itself into the cell or organism's reproductive DNA so that a copy of the virus is then present in every cell of that organism's offspring. This is potentially an even more effective way to "go viral."

The human genome has large pieces of many different retroviruses. The junk-DNA crowd (a group that insists 97 percent of your DNA is "junk"; see chapter 31) has touted this as evidence that our genomes are littered with just so much "evolutionary garbage." That might make perfect sense if cells were helpless victims of chance and necessity. But, as we've been seeing from chapter 11 onward, they are not (655).

One might assume these viruses are purely destructive, accumulating in your DNA like rusty engine parts in a junkyard. But retrovirus code sequences turn out to be symbiotic and very useful to creatures that survive the initial viral attack. Organisms edit their genomes, using Transposition to harness the new code found in the virus. Cells engineer new capabilities through adaptive mutations of this new code. We have strong evidence that this happened multiple times in the development of mammals.

RETROVIRUS

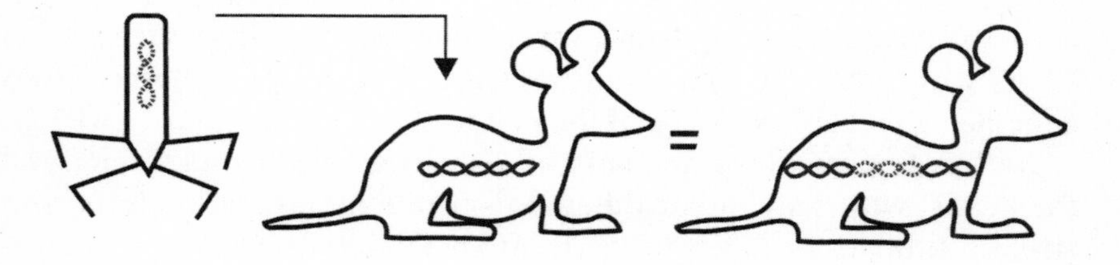

Regular viruses hijack cells and turn them into virus-manufacturing factories. Retroviruses are different; they insert their DNA into the genome so that it replicates with the host. This can kill the host, but hosts that survive now have additional DNA, which becomes useful for later adaptations.

For example, in mammalian placentas there is a membrane one cell thick called the *syncytium*. It mediates transfer of waste and nutrients between mother and child. Genome research indicates that the instructions to build the syncytium first came from code in the envelope gene of the HERV-W human endogenous retrovirus. Also, so far as we can tell, different versions of the syncytium were constructed in mammals three different times from entirely different viruses.* (652)

Is Symbiogenesis Just a Theory? Or Has It Been Produced in the Lab?

Unlike Darwinian theories of gradual evolution over eons of time, symbiotic cell mergers have indeed been observed in the lab. In 1939, German botanist Eugen Thomas synthesized lichen from its constituent parts (677); later, others did similar experiments (601).

Dr. Kwang Jeon, a professor at the University of Tennessee, did an experiment where tens of thousands of bacteria took up residence inside *Amoeba proteus* organisms. A fierce parasitic attack ensued, killing almost all the amoeba. But in the space of a year, amoeba and bacteria entered into symbiosis. Both modified expression of their genes as necessary, to support the mutual dependence (624, 636, 653, 652).

Jeon learned how to reliably trigger symbiotic cell mergers between amoeba and bacteria. It took 200 generations, about 18 months, for the cells to become fully interdependent. After that, removal of either symbiotic partner proved fatal to both (625).

By doing these experiments, researchers like Eugen Thomas and Kwang Jeon, rather than merely theorizing or fighting about evolution, proved it in the lab in real time. They transformed evolution from the 1.0 version—a historical science where people make assertions based on fossils, guesswork, and anecdotal evidence—to a 2.0 version, an empirical discipline where we discover how to generate new species at will. These are the real heroes of evolution.

---

* I only briefly mention retroviruses here, but they are described in fascinating detail in Frank Ryan's eye-opening book *Virolution* (652).

Cooperation, Not Survival of the Fittest

Symbiogenesis was first described in detail by Russian scientist Boris Mikhaylovich Kozo-Polyansky in his 1926 book *Symbiogenesis and the Origin of Species*, which received widespread acceptance in Russia but, because it wasn't translated into English for 84 years, was almost lost entirely.

Fortunately, leading American biologist Dr. Lynn Margulis—who in her lifetime received the William Procter Prize for Scientific Achievement, the Darwin-Wallace medal from the Linnean Society of London, and the National Medal of Science—excavated the theory of Symbiogenesis, extended it, and introduced it to the West.

In her 2003 book, *Acquiring Genomes: A Theory of the Origins of Species* (637), Dr. Margulis argued that Symbiogenesis is a primary driver of evolution. She said there's no good laboratory evidence that random mutations cause inherited variations to occur. Instead, new organelles, bodies, organs, and species arise from Symbiogenesis.

Obviously, Dr. Margulis was no Darwin groupie. While the reigning 1.0 version of evolution emphasizes competition as the primary force, Margulis focused on harmony and cooperation. She insisted that history will ultimately judge Darwinian evolution as having been "a minor 20th century religious sect within the sprawling religious persuasion of Anglo-Saxon biology" (635).

Darwin's principle of natural selection still applies, of course—once new organelles, bodies, organs, and species exist! But without these *other* powerful creative forces in DNA (Symbiogenesis as well as the other parts of the Evolution 2.0 Swiss Army Knife we discuss in this book), natural selection gets nothing new or useful to operate on. Without the Swiss Army Knife, nothing evolves.

And please don't miss the fact that the initial symbiotic merger occurs in a single step! *This means that, in Symbiogenesis, natural selection doesn't necessarily have any direct involvement at all.* Natural selection only acts as a platform for the benefit to proliferate once it exists.

Nature is so often depicted as cruel and merciless in its bitter and unrelenting struggle. But when you actually spend time in nature—when you slow down enough to watch what is going on around you—you witness fabulous, intricate interdependence. Grass keeps soil from eroding. Bees and flowers engage in a dance with each other. Bacteria colonies nourish plant roots. Worms tunnel through the soil, opening pathways for air, water, and nutrients.

Big fish get their mouths cleaned by "cleaner fish." The cleaners get lunch, and the clients get rid of that film in their mouths.

When a predator approaches the pied flycatcher bird, it screeches loudly to alert others. It's taking a risk because predators can take notice and kill the flycatcher. But it's worth the tradeoff because the more birds join in, the better the chances that the enemy will retreat. (249)

Nature's depiction as ruthless and bloody—recall those wildlife shows where you watch leopards hunt and eat gazelles—is overstated. Cooperation and symbiosis are so ever-present we tend to look right past them and only notice the competition.

Darwin's version of evolution was admirable for its time. But as you can see now, his work is largely outdated. Worse yet, Darwin's theory experienced an immense setback in the 1930s with the Neo-Darwinian assumption of random copying errors. Darwinism has progressed from benign to destructive as Darwin's successors have consistently suppressed and torpedoed superior models.

BULLET POINT SUMMARY:

- Neo-Darwinism says Random Mutation + Natural Selection + Time = Evolution.
- Random mutation is noise. Noise destroys.
- Cells rearrange DNA according to precise rules (Transposition).
- Cells exchange DNA with other cells (Horizontal Gene Transfer).
- Cells communicate with each other and edit their own genomes with incredibly sophisticated language.
- Cells switch code on and off for themselves and their progeny (Epigenetics).
- Cells merge and cooperate (Symbiogenesis).

CHAPTER 16

Blade #5: Genome Duplication—Evolution at Lightning Speed

Leading a double life
Can it be wrong when you know that it's right?

—STYX

IN 1972, Stephen Jay Gould, a celebrated paleontologist at Harvard University, and Niles Eldredge, curator of invertebrates at the American Museum of Natural History, jolted the Darwinian establishment with a new theory called Punctuated Equilibrium (614).

Punctuated Equilibrium says that evolutionary progress hovers near zero for long periods of time, then suddenly makes huge forward leaps. The theory explained why the fossil record showed species remaining stable for millions of years, then suddenly being joined by a new species, almost overnight.* It quickly became a source of bitter controversy among evolutionists. (669)

* In these examples, "overnight" means, say, 30 million years. A long night indeed, but a mere instant in Earth's history. Hey, it might have even taken less time than that, but it's hard to parse smaller spans of time than that in the fossil record.

Achilles' Heel of Darwinism

Old-school "gradualists" (people who deny evolution can happen quickly, because they believe random mutations and natural selection do most of the work) were furious with Gould and Eldredge. To suggest that evolution did not occur gradually posed all kinds of new problems for evolutionary theory. Gradualists considered them traitors of sorts. Richard Dawkins* was enraged with Gould for committing scientific treason:

> The extreme Gouldian view... is radically different from and utterly incompatible with the standard neo-Darwinian model. It also... has implications which, once they are spelled out, anybody can see are absurd... For a new body plan—a new phylum—to spring into existence, what actually has to happen on the ground is that a child is born which suddenly, out of the blue, is as different from its parents as a snail is from an earthworm. No zoologist who thinks through the implications, not even the most ardent saltationist [evolutionist who believes in quantum leaps], has ever supported any such notion. (113)

Gould and his detractors alike chafed at the fact that Creationists also used his theory as a weapon against evolution itself. Darwin predicted intermediate forms; to Creationists, no intermediate forms meant no evolution.

But Gould was only pointing out that the intermediate forms and gradual progress that Darwin predicted are poorly supported by the fossil record. It doesn't have enough transitional forms—some, but not many. This fact is well known. It has always been the Achilles' heel of Darwinism. It's even sometimes called "the trade secret of paleontology" (616).

Darwin knew it himself back in 1859. He just didn't realize the extent of the problem. He wrote in his original *Origin of Species*, "Why then is not every geological formation and every stratum full of such intermediate links? Geology assuredly does not reveal any such finely graduated organic

---

* Dawkins' popularity has declined for a variety of reasons we won't explore here. Nevertheless his influence on modern evolutionary thinking is tremendous, not just in popular culture but in scientific literature. It is impossible to talk about evolution without referencing Dawkins.

chain; and this, perhaps, is the most obvious and gravest objection which can be urged against my theory" (108).

The fossil record shows the first life forms appearing more than 3 billion years ago. Then, for more than 2 billion years, you see only simple life forms, and only in the ocean (237). Then, suddenly, 540 million years ago, the fossil record shows a quantum leap of unprecedented proportions, known as the *Cambrian explosion* (209), in which we witness a massive expansion of life forms. By the end of the Cambrian explosion—500 million years ago—some 40 new phyla (major animal groups like arthropods and chordates) had appeared (209). The majority of this progress clearly took place in a relatively short span of time. Branches of the tree of life pop up all over the place in quick succession. Not exactly evidence of Charles Darwin's theory of "gradualism"!

The Cambrian explosion was a wide-open door for Creationists' derision of evolution. Scientists and historians alike acknowledge that it challenged Darwin's conception of a gradual evolution (617, 634, 619).

Gould and Eldredge gave us a name for this: Punctuated Equilibrium. For decades there was no agreed-upon understanding of how this actually happened, only acceptance from factions within the evolutionary community that it did happen. It's been a huge missing puzzle piece in the story of evolution.

It's pretty hard to imagine how a slow accumulation of random copying errors could have so little effect for long spans of time, and then somehow coalesce into so many new species, genera, and phyla so quickly. Of course by now you know about Symbiogenesis, which can happen very fast. But that's not the only potential explanation for Punctuated Equilibrium.

By Leaps and Bounds—Patterns in Science and Technology

We see Punctuated Equilibrium in the progress of science itself. For centuries Newton's theories of physics (935) held steady, then, over the course of less than 20 years, Einstein's theory of relativity and the development of quantum mechanics superseded them.

Likewise, the standard for TV signals remained unchanged for decades, then HDTV was introduced. In just a few years, millions of people embraced this quantum leap in picture quality.

Barbara McClintock's discoveries about Transposition, which I discussed in chapter 11, begin to explain Punctuated Equilibrium very well. If organisms can radically reorganize their own DNA in one generation, then evolution might happen very rapidly... but only under certain conditions.

Which brings us to Genome Duplication.

Genome Duplication

Punctuated Equilibrium wasn't the only evolutionary theory that gained traction in the 1970s. There was another, called Ohno's 2R hypothesis (648, 626), aka Genome Multiplication or Whole Genome Duplication. It's even more surprising than what McClintock discovered with Transposition. It's blade #5 of Evolution 2.0's Swiss Army Knife.

Susumu Ohno was a Japanese American geneticist and evolutionary biologist, a seminal researcher in the field of molecular evolution. He wrote a classic book called *Evolution by Gene Duplication* (648). His 2R hypothesis suggested that the genomes of the early vertebrate lineage underwent one or more complete Genome Duplications (also known as "Genome Doublings"). These duplications fueled sudden, radical transformations of body plans.

2R stands for "2 Rounds" of doubling. The first, Ohno suggested, was at the origin of vertebrates, and the second at the origin of jawed vertebrates. It explains how, in a few exceedingly rare events, the genetic chassis for new species could be built in a single generation.

For instance, careful study of genome sequences suggest the hagfish originated exactly this way about 500 million years ago. This appears to have happened through interspecies Hybridization—the mating of two species—in the hagfish's case, two species of sea squirt. Here's how.

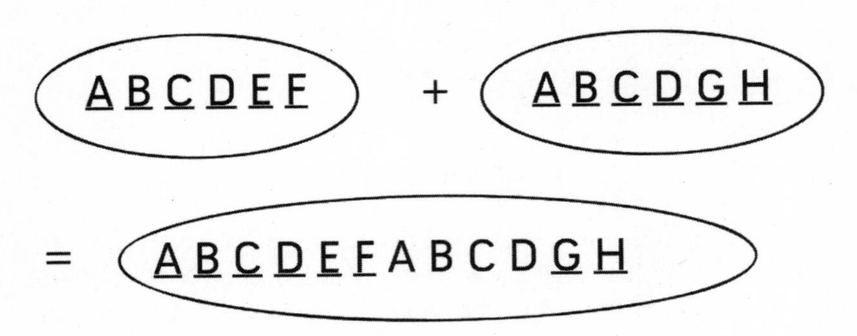

Hybridization is when two species merge to form a new species. This doubles the number of chromosomes, dramatically increasing the genetic material available. A merger is followed by "hybrid dysgenesis," a process in which the cell rearranges the new DNA strand and discards coding sequences that it doesn't need. The underlined letters are DNA sequences that are "switched on" epigenetically. Non-underlined letters are switched off.

Hybrids Make New Species

Darwin gave us the popular conception that new species gradually emerge over time. But dogs all by themselves will never become anything other than dogs unless they mate with some other animal that is not a dog. Breeders and geneticists have known this experimentally for hundreds of years (673). Hybridization is the most reliable way to produce new species. One set of parents can produce offspring with new genetic features, who then mate and introduce a new species to the population.

When different species mate, the offspring with its doubled chromosomes is usually sterile (male donkey + female horse = sterile mule)—but not always. In plants and some animals, like salmon and salamanders (607), the result of different species mating is that their genome doubles. Instead of having a combination of the mother's and father's genes in a single set of chromosomes, the child inherits double chromosomes—a full set from its mother, and a full set from its father. This is called *diploidy*.

In the offspring's doubled DNA, many of the code sequences are turned off, but remain present. The process of joining the two DNA strands together also, in rare matings, provokes rearrangements through Transposition. This sudden rearranging is called *hybrid dysgenesis*, and it can provoke sudden new and useful features its parents never had. Bread wheat, for example, came from blending emmer wheats with goat grass, a noxious weed, meaning one of the world's most popular crops came from blending a moderately useful crop with a useless pest (673).

An article in *ScienceDaily* titled "Two Species Fused to Give Rise to Plant Pest a Few Hundred Years Ago" (679) documents how a fungus originated from a hybrid 380 generations in the past. The researchers identified specific transpositions in the hybrid's doubled genome that stabilized into a brand new, stable species.

Doubled DNA boasts twice as much storage space as before, and organisms make greater use of that space through Transposition. In fact, the organism uses the duplicated genome as spare "lab space" for experimentation (628)! This is how two sea squirt species gave rise to the hagfish species in what appears to have been a very short period of time.

The other day I happened to be talking to the CEO of a seafood wholesale company. A friend asked him, "In all your travels, what's the most disgusting seafood you've ever eaten?"

"Hagfish," he replied. "Koreans like them for some reason. In the business we call them slime eels. When you squeeze them, they protect themselves by producing slime from their skin. It enables them to get away when predators try to catch them. We got a hagfish shipment once and the slime on the dock was so bad our forklifts wouldn't work anymore."

A hagfish is a jawless vertebrate about 50 centimeters long that feeds on the bottom of the ocean. A sea squirt is a sea creature with no backbone. It's about 5 centimeters long and is usually found beneath the sand in shallow parts of tropical seas. Water is filtered inside its sack-shaped body. Sea squirt larvae swim free; eventually they lose their tail and ability to move, and as adults attach themselves to something hard in the ocean.

At first glance, hagfish and sea squirts seem to have very little in common. But through Genome Duplication, the hybrid of two invertebrate sea squirt species transformed into the vertebrate hagfish, in a small number of generations. Multiplication of the genome accomplished dramatic change in a short period of time.

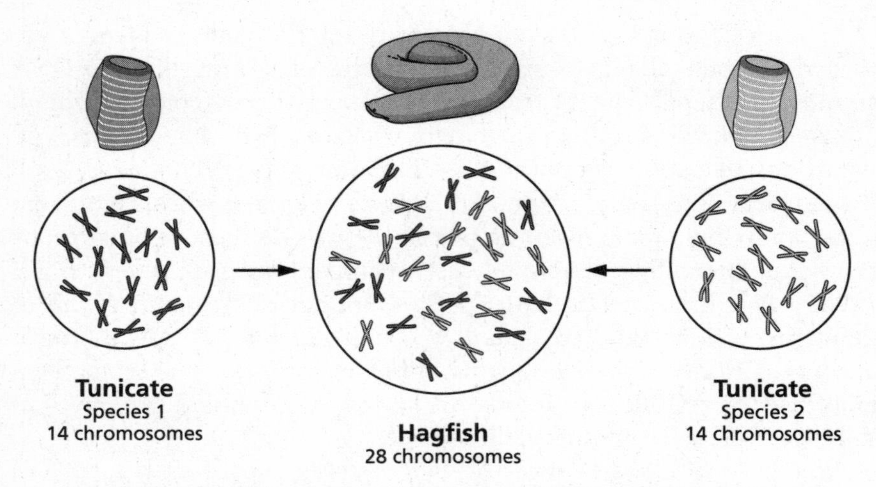

The genetic platform for a hagfish was built in one generation when two species of tunicate (sea squirt) merged to create a third species. This new creature had twice as many chromosomes through Genome Duplication. The new genetic material gave other mechanisms like Transposition freedom to develop new features.

How short? Ten generations? A hundred? A thousand? The length of this period is unclear. But by Darwinian standards, it certainly was *not* gradual! However long it took for things to settle down, the genome itself doubled in one generation.

Further evolution was then possible because a higher capacity for information was present in the new creature's genome; the new rearranged genes were expressed differently. And the hagfish meets the definition of "new species": Hagfish can breed with each other but cannot breed with sea squirts, just as wheat plants can breed with each other but not with the weeds that produced them.

The sea-squirt-to-hagfish transition was the first key round of doubling. Then, a short time later (only about 50 million years), a second Genome Duplication between two species occurred. Now the genome was four times its original size. The resulting creature was the ancestor of the world's first jawed vertebrate (659). This was an early ancestor of bony fish, reptiles, amphibians, birds, and mammals.

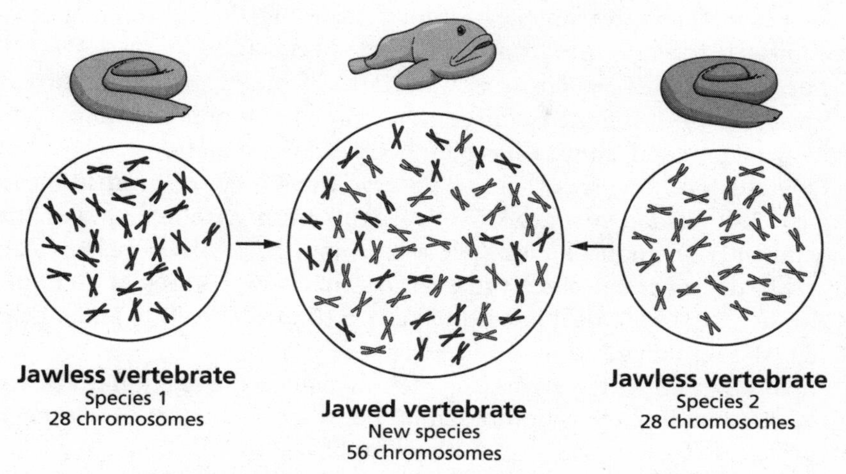

Jawed vertebrates came from a second merger of two species, which again doubled the number of chromosomes. Members of the new species could breed with each other because they both have 56 chromosomes but could not breed with the old species, which still had 28. New lineages can appear alongside old ones, and both will coexist.

Usually genomes double, but sometimes they triple through abnormal cell division. Watermelons have three genomes; wheat has six.

No Case for Randomness

Again, DNA evidence seems to indicate that the genetic capacity to engineer the new jaw appeared in a *single generation.* It's unclear if the jaw itself appeared in one or more generations. It's likely that the 2R event only created the conditions for the jaw to form some time later. In any case, we directly infer this from the data because very similar code patterns are found in two places in plant and animal genomes after one doubling; then in four places after another doubling.

These findings build our confidence that progress is driven by adaptive mutations—modular reengineering of genes and chromosomes—not by random copying errors. James Shapiro, who you'll remember from our discussion of Transposition, suggests that Hybridization is more likely after an environmental crisis like an earthquake or forest fire, because creatures unable to find a mate within the same species would settle for a partner from a different species (659).

For years the 2R Hypothesis was hotly disputed; however, during the last decade it's gained considerable support as more genomes are sequenced. Sequenced data has matched Ohno's predictions.

Obviously no one was there to observe the sea squirt's transition to hagfish 500 million years ago. We infer this from genetic data. But the general phenomenon of new species through Hybridization has been superbly documented since the 19th century—with cases of Genome Multiplication in plants like wheat and rice (673), butterflies and moths (632), and donkeys and mules (243).

In some groups of vertebrates, such as fish of the salmon and carp families (including the zebrafish, a popular research animal), it has been suggested that there was yet another duplication ("3R"), so they have eight times as many genes as the original invertebrates (641).

A Simple Illustration of Genome Duplication, Using English

Let me illustrate how this looks at the code level. Since very few of us are equipped with the capacity to read strands of DNA, I'll use English.

Here's a simple illustration of Genome Duplication: two nearly identical sentences in a row, then modified by gene substitution. Bold indicates Transposition, and gray indicates modifications by epigenetic systems:

1. The quick brown fox jumps over the lazy dog. The slow golden fox jumps over the crazy dog.
2. The quick brown fox jumps over the lazy dog. The slow fox jumps over the crazy, **golden** dog.
3. The **slow** brown fox jumps over the lazy dog. The **quick** fox jumps over the crazy, golden dog.

Now let's double it again—4R instead of 2R (the third and fourth sentences are from yet another "species"):

4. The slow brown fox jumps over the lazy dog. The quick gold fox jumps over the crazy, golden dog. The slow red fox leaps over the lazy dog. And the fast red fox jumps over the big black dog.

Now, after a few more stages of evolution (copying and transposing some of the words):

5. The slow brown fox and the red fox jumps over the lazy golden dog,. And the quick gold fox jumps over the crazy dog,. The slow red fox leaps over the lazy dog. and the fast red fox jumps over the big black dog.

Last, hybrid dysgenesis not only repurposes the new genetic material but also deletes unwanted genes altogether:

6. The slow brown fox and the red fox jump over the lazy golden dog, the crazy dog, and the black dog.

In English, we make this happen by (1) knowing the rules of grammar, and (2) choosing in advance how we want to shape the story. The starting point is the structure in our root sentence, which in this case is:

[adjective] [noun] [verb] [adjective] [noun]

As we evolved this paragraph, we changed one fox to multiple foxes and changed "jumps" to "jump." Don't forget that grayed-out codes are still in the genome, ready perhaps to be combined in ways they've never been before.

This example is highly simplified, I admit, but it does convey a fairly accurate sense of what the cell does in the process of evolving its own DNA. Cells make precisely these kinds of substitutions. They select what's useful to them in the same way we selected what parts of the sentence we wanted to keep.

Just as "brown" refers to "fox," genes are interconnected and refer to each other. Genome Duplication and subsequent refinements really do resemble intelligent substitution of nouns, verbs, adverbs, and adjectives. As you'll see in chapter 19, DNA really is a language! That's why it's possible to generate new species by transposing pieces of duplicated genomes.

A Natural Function of DNA, Not an Accident

Some might be tempted to suppose that Genome Multiplication is itself some sort of copying error or rare freak accident. But it's a well-known adaptive mutation that occurs routinely in many plants. Hybridization reliably produces new species of plants, and plant growers make hybrids at will.

What kind of software makes a copy of its own code, splices it to the original, and—with rapid, on-the-fly changes—generates an entirely new program with brand new features that never existed before? No human programmer has ever written software that does anything even close to this.

Could randomness do it?

No chance.

Revisiting Fruit Flies

In chapter 4, you saw how researchers in the early 20th century hoped to accelerate evolution by exposing fruit flies to radiation, expecting to trigger random mutations in their genomes. They hoped to find some optimum level of radiation at which the flies would evolve at maximum speed.

These experiments didn't produce the expected outcome. The only reason a few of these experiments showed even a hint of success was because cellular repair mechanisms fixed the broken ends of the DNA. In those cases, the *repair* (performed by a nonrandom, goal-seeking cellular system) would occasionally confer a desirable trait. That's what happened to Barbara McClintock's corn plants, too.

Based on what we've learned since, the fast way to provoke fruit flies to generate new species would be to breed them with a similar but separate species of fly, forcing Genome Duplication. Hybrids have a high rate of failure, but when they work, they immediately produce a new species. Genome Duplication activates epigenetic activity and Transposition events, which accelerate the flies' natural adaptive machinery.

Alfred Sturtevant was an American geneticist who constructed the first genetic map of a chromosome in 1913 and received the National Medal of Science in 1967. He began experiments of this very kind with fruit flies in the 1920s, and although they proved difficult (almost all of his flies' offspring were sterile), this eventually proved to be an excellent line of inquiry. These experiments show that radical adaptation is triggered by specific types of events. Hybridization is a major factor.

In his paper "Ninety Years of *Drosophila melanogaster* Hybrids" (603), Daniel Barbash, a professor of genetics at Cornell University, describes fruitful experiments inspired by Sturtevant's original work. He reports, "Many behavioral, ecological, Population Genetics, and gene expression differences between these species have since been discovered." These experiments spawned interesting variants that radiation experiments failed to produce.

Hybridization triggers the genome to be rearranged according to rules that we are only just beginning to understand in detail, but one thing is clear: Radiation alone would never produce evolution, even if you had millions of years at your disposal.

We have witnessed many genome duplication events and new species in the lab. I do admit that none we have witnessed there are as dramatic as the transition from sea squirt to hagfish. This necessarily would have required utterly remarkable cellular engineering, including construction of several new body parts. I believe that as we study Hybridization, Evolution 2.0's Swiss Army Knife will further impress us.

Origin of Species?

Darwin himself recognized Hybridization as a means of creating new species, and he referred to it a number of times in his famous book. He didn't emphasize it, though, and there's no indication he grasped its true significance.

The Neo-Darwinian synthesis in the 20th century granted it even less credit. Fortunately it did begin to receive more attention in the 1950s with the work of Ledyard Stebbins at the University of California–Davis (673). Stebbins was one of the leading botanists of the 20th century. His name is a household word in genetics.

In popular culture, Charles Darwin is believed to have discovered where new species come from. In actual reality, he did not. Hybridization is the *only* process mentioned in Darwin's book *On the Origin of Species* that is directly observed to originate a new species. And... it was known long before Darwin! Experiments that created new species before Darwin were hybrids, and post-Darwinians created new species through Symbiogenesis.

The Five Blades of the Evolution 2.0 Swiss Army Knife

If we rank these blades from most gradual to most sudden, the list looks like this:

1. Epigenetics
2. Transposition
3. Horizontal Gene Transfer
4. Hybridization
5. Symbiogenesis

Not surprisingly, the first events in the list are far more common than the last.

And what happens when we combine Transposition (cells rearranging their own DNA), Horizontal Transfer (cells exchanging DNA), Epigenetics (organisms passing acquired traits to offspring through cells switching DNA sequences on and off), and Symbiogenesis (organisms merging together) with Genome Duplication (two species merging to form a third)?

With these five blades of the knife, we can in principle get from any one spot on the tree of life to any other. Also remember the retrovirus—a symbiotic corkscrew, which leads to organisms borrowing code from code inserted into the virus' own DNA. Finally we're beginning to form a reasonable sketch of how life proliferated on Earth—how life has managed to fill every niche with amazing diversity.

Most important, what we've learned about Evolution 2.0 is:

- It's not slow or gradual; it's fast.
- It's not accidental; it's organized.
- It's not purposeless; it's adaptive.
- Natural selection isn't the star of the show; Natural Genetic Engineering is.

Oxford professor Denis Noble is president of the International Union of Physiological Sciences and author of *The Music of Life* (644). Life, Noble asserts, is a kind of music, a symphonic interplay between genes, cells, organs, body, and environment. Noble has written that "all the central assumptions of the Modern Synthesis (often also called Neo-Darwinism) have been disproved. Moreover, they have been disproved in ways that raise the tantalizing prospect of a totally new synthesis." (645)

ONLINE SUPPLEMENT

Dr. Denis Noble of Oxford on post-Darwinian, post-Dawkins evolution

www.cosmicfingerprints.com/supplement

Saying "Mammals evolved by natural selection" is like saying "The Seahawks made it to the Super Bowl by winning the playoffs." That might be a perfectly satisfactory answer *to a person who has no interest whatsoever in the minutiae of football games*, but no true football enthusiast would be content with it. Sports fans demand complete explanations, from recruiting and injuries and talents of star players, to the coach's defensive and offensive strategies; from fouls and referee calls to stats and specific moves in the team's signature plays.

If sports fans demand detailed explanations for the outcome of a championship football game, why should science fans insist on anything less for evolution?

"Evolution through natural selection" (the 1.0 version) left out all the interesting stuff. Every bit of this remarkable process owes its existence to an exquisite array of signals from the environment, communication between cells, and the ingenuity of cells themselves. It's so fantastic that few even dared to imagine it, but it's been right under our noses all along.

Finally: Evolution 2.0

The five blades of the Swiss Army Knife bring us to a brand new understanding of evolution—*Evolution 2.0*. Evolution 2.0 is defined as *the cell's capacity to adapt and to generate new features and new species by engineering its own genetics in real time.*

#EVOLUTION IN 140 CHARACTERS OR LESS

Genes switch on, switch off, rearrange, and exchange. Hybrids double; viruses hijack; cells merge; winners emerge.

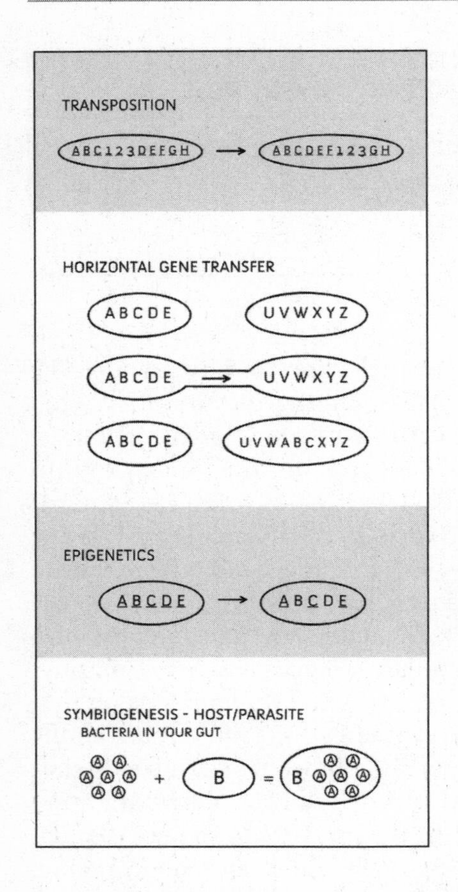

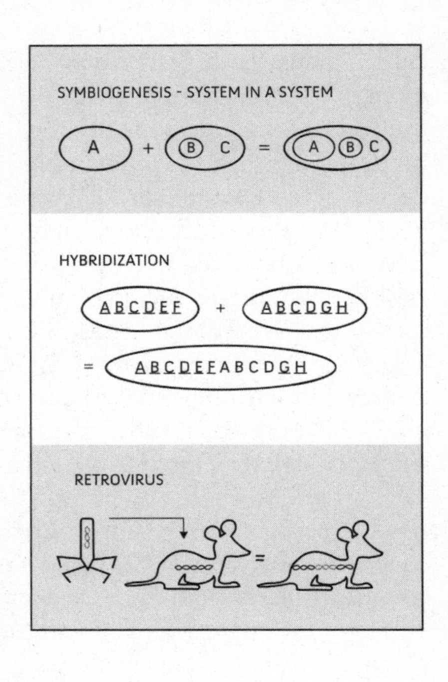

Evolution 2.0 is driven by a Swiss Army Knife of multiple cell-engineered systems. Transposition rearranges segments of DNA. Horizontal Gene Transfer shuttles DNA between cells. Epigenetics switches genes on and off (underlined letters = "on"), allowing acquired adaptations to be passed to offspring. Symbiogenesis merges cells to form new organisms with new capabilities. Hybridization doubles chromosomes, so species 1 + species 2 = new species. Retroviruses inject new DNA sequences into hosts. Last, natural selection sorts winners from losers. Each of these mechanisms occurs in real time and has been produced in the lab.

many people, "evolution" has become a four-letter word, a stick wielded by anti-theists in mannerless debates about science and religion. It drips with all kinds of cultural and political baggage. As you see in the chart opposite, Evolution 2.0 is different. It sheds the negative Darwinian connotations of a blind, bloody battle of luck and selection for an endlessly fascinating, highly directional Swiss Army Knife.

Evolution's Swiss Army Knife

Issue	Evolution 1.0 (Neo-Darwinism)	Evolution 2.0
Origin of Life	Presumed to have emerged from random chemical processes	Information theory says codes require a designer, or an undiscovered natural process that generates codes
Speed	Gradual	In real time
Sources of Novelty	Random copying errors; natural selection is the hero	Transposition, Horizontal Transfer, Epigenetics, Hybridization, Symbiogenesis; natural selection was overrated (evolutionary steps occur *before* natural selection, not after)
Scientific Status	Randomness impossible to prove; much of the evidence is anecdotal, not empirical; millions of years too long to test	Demonstrated in 70-plus years of documented live lab experiments
Implications for Humanity	Chance, luck, and "blind pitiless indifference" of an uncaring universe; social Darwinism	Profoundly directional, cooperative process that invites us to humble ourselves and study with care
Implications for Science & Technology	Humans are smarter than nature, so we must now begin to direct our own evolution	Nature is far wiser than we are, suggesting caution; cell research promises tremendous breakthroughs in medicine and engineering
Implications for Spirituality	Religion is a myth, a way for "holy men" to wield power over the masses	Science points to something beyond itself, far greater than us or the universe

BULLET POINT SUMMARY:

- Neo-Darwinism says Random Mutation + Natural Selection + Time = Evolution.
- Random mutation is noise. Noise destroys.
- Cells rearrange DNA according to precise rules (Transposition).
- Cells exchange DNA with other cells (Horizontal Gene Transfer).
- Cells communicate with each other and edit their own genomes with incredibly sophisticated language.
- Cells switch code on and off for themselves and their progeny (Epigenetics).
- Cells merge and cooperate (Symbiogenesis).
- Species 1 + Species 2 = New Species (Hybridization). We know organisms rapidly adapt because scientists produce new species in the lab every day.
- #Evolution in 140 characters or less: Genes switch on, switch off, rearrange, and exchange. Hybrids double; viruses hijack; cells merge; winners emerge.
- Adaptive Mutation + Natural Selection + Time = Evolution 2.0.

CHAPTER 17

Why Is Neither Side Telling You the Whole Story?

Stars, hide your fires;
Let not light see my black and deep desires.

—William Shakespeare

IN 2009, the famous atheist Richard Dawkins published his thick, best-selling book *The Greatest Show on Earth*. In it, he states that evolution is driven by random changes in genes. It is worth noting that in all of 450 pages of *The Greatest Show on Earth* . . .

- Symbiogenesis is never mentioned.
- Horizontal Gene Transfer is briefly touched on once, downplayed and presented as scarcely ever crossing from one species to another.
- Epigenetics gets one tiny footnote in chapter 8. He breezily shrugs it off as a "modest buzzword" and "confused theory that will enjoy 15 minutes of fame." (At the time of this writing, "Epigenetics" is a major focus in genomics and appears 129,000 times in Google Scholar. The number of entries has doubled in the last two years—clearly a hot field of research.)
- Transposition is never mentioned.
- Genome Duplication is never mentioned.

Why didn't Dawkins grant so much as three pages to the five best-documented mechanisms of evolution? Why does he act as though the last 50 years of microbiology and billions of dollars of research never happened? Oxford University's former "Professor of the Public Understanding of Science" wrote one of the most popular evolution books of the last decade, for which he received large advances and rode huge waves of media publicity.

So why isn't he disclosing this?

On the other side of the fence, Stephen Meyer, in his pro–Intelligent Design book *Darwin's Doubt*, makes an eerily identical set of omissions (130). Epigenetics gets decent airtime, but there's no explanation of Lynn Margulis' work on Symbiogenesis. Barbara McClintock, Transposition, Horizontal Gene Transfer, and Genome Duplication are touched on only briefly, mostly in footnotes.

New Genetic Information?

On pages 332 to 335, Meyer is gracious to James Shapiro's model of Natural Genetic Engineering, but in the end Meyer takes as a foregone conclusion that Natural Genetic Engineering is still incapable of producing the Cambrian explosion. A frequent Creationist and ID claim is that there is no known observable process by which new information can be added to the genetic code of an organism (310).

Yet experimental evidence does show that Natural Genetic Engineering is capable of producing major evolutionary events and that targeted evolutionary changes respond to hundreds of inputs from the environment. According to information theory, if an organism editing its own genome can *choose* between a "1" and a "0," then it is in fact creating new information. The decisions organisms make as they apply the Swiss Army Knife create genetic information.

Meyer and his pro-Darwin opponents are making identical, equal, and yet opposite, mistakes. Both move evolutionary steps out of the realm of scientific discovery and into ineffable mystery, so round and round it goes. Thus the deadlock between Darwin and Design. Both sides have missed the biggest story in the history of science.

In his famous book *The Blind Watchmaker*, Dawkins said, "It is almost as if the human brain were specifically designed to misunderstand Darwinism, and to find it hard to believe" (111).

In the movie *Expelled,* Dawkins is pressed for an explanation of the Origin of Life. He says, "It could be that at some earlier time, somewhere in the universe, a civilization evolved by probably some kind of Darwinian means to a very, very high level of technology—and designed a form of life that they seeded onto perhaps this planet…And I suppose it's possible that you might find evidence for that if you look at the details of biochemistry, molecular biology, you might find a signature of some sort of designer."

Yes, even the world's most famous evolutionist and atheist, caught off guard by a well-worded question, admits living things may bear the signature of design.

The Miracle of Evolution 2.0

Nothing we presently know in pure physics or chemistry explains the origin of these cellular engineering capabilities. We don't know how cells make choices. If the universe booted up the first cell without the action of a designer, then the universe itself must possess directional qualities that nobody yet comprehends.

Humans make machines. To date, we've never made self-replicating machines. The closest we've come is computer viruses. But what if someone built a computer that reproduced other physical computers? What if someone wrote a computer program that got better and better with time, all by itself?

What if Bill Gates started with DOS and we got the latest version of Windows without any human programmer ever having to write a line of code? That would be pretty impressive, wouldn't it?

We all know that human-made machines don't evolve all by themselves; left to themselves, our cars and computers and PlayStations degrade and crash and break.

So…which is harder:

1. Building a machine that can only make other similar machines?

or

2. Building a machine that can make other machines, which exponentially improve with time?

Here's another version of the same question. Which is harder:

1. Designing a zebra?

or

2. Designing a cell that builds an ecosystem for a zebra to live in—then builds the zebra, too?

EVOLUTION 2.0 AND THE LANGUAGE OF CELLS

CHAPTER 18

Curious George and the Blog Spam Theory of Evolution

"Shush dear, don't have a fuss. I'll have your spam. I love it!
I'm having spam, spam, spam, spam, spam, spam,
spam, baked beans,
spam, spam, spam, and spam!"
"Spam! Spam! Spam! Spam! Lovely spam! Wonderful spam!"

—Monty Python

I ALWAYS LIKED CURIOUS GEORGE BOOKS. My favorite one is *Curious George Learns the Alphabet*, and my favorite part of it is where the Man with the Yellow Hat is reviewing George's new words (see next page).

Cheeky as he is, the Man with the Yellow Hat makes a point very much pertinent to evolutionary biology. To illustrate, allow me to take you on a side trip into a different kind of evolution: the evolution of... internet spam. You'll see interesting parallels between the two evolving worlds of digital information and cellular information. And yes, it might even remind you of Curious George and his Dalg, Glidj, and Blimlimlim.

In evolutionary theory, "random" in regard to mutation has generally meant one of two things.

"Let me see," said the man. "Ball–Milk–Cake–Ham–Jam–Egg–Lime–Feed–Kid–that's very good.

But what on earth is a Dalg or a Glidj or a Blimlimlim? There are no such things. Just ANY letters do not make words, George.

Well, let's look at some new letters now."

1. **Truly random copying errors in DNA.**

   Example: "The quick b%own fox jumps over the l3zy dog" where the % and the 3 are random letter substitutions.

   We could call that the Email Spam Theory of Evolution, as I shall explain shortly.

2. **DNA changes that have no forward-looking goal or purpose (which Neo-Darwinism assumes to be all of them).** In the words of the famous old-school Darwinist Jerry Coyne: "Mutations occur regardless of whether they would be useful to the individual."

   Consider this example of a sentence with real words, in which the words are arranged randomly and purposelessly:

   Lazy dog brown over quick jumps fox the

You could call that the Blog Spam Theory of Evolution.

The Email Spam Theory of Evolution

In the 20th century, spam was mostly just email. Bots harvested email addresses from websites. If your email appeared anywhere online, the spammers got their hands on it.

Have you ever gotten a spam email that looked like this?

> Subject: foes called him
> TR ADERS WATCH OUT
> SBNS ROCKS!
> WAT CH SBNS TRADE ON THURSDAY OCT 12!
> Trade A lert: THURS DAY, October 12, 2006 Company Name:
> SHALLBETTER INDS INC (Other OTC:SBNS.PK) Price: $0.95
> SYMB0L: SBNS.PK 5-day Targ et: $10
> NEWS
> - Shallbetter Industries, Inc. Provides Geological Information
> Relating To Initial Resource Property In Mongolia READ MORE
> ONLINE!
> zb6
> ~~~~ (100)

You immediately notice random characters and misspelled words. These random bits of text are called *hash busters*. They fool spam filters, which look for specific text patterns and common spam trigger words.

These messages don't work on humans very well, as we all know. The only reason they work on humans at all is because when we see misspelled words, our minds automatically correct them without thinking about it.

Hash busters, notably, never create emails that are superior to genuine human-made emails.

The Blog Spam Theory of Evolution

As the web evolved, so did spam.

Google introduced its advertising program, AdWords, in 2002. Not long after that they launched a new program called AdSense that pays webmasters to put Google ads on their site. It didn't take long for spammers to find a way to profit from this, too.

Google charges you as an advertiser every time someone clicks on your ad. They might charge you 10 cents, a dollar, or 10 dollars, depending on the auction price. (A tiny handful of hyper-competitive keywords, such as "austin texas driving while intoxicated," can approach 100 dollars per click.) In this new program, you could host Google ads on your site, and every time people clicked on them, Google would split half to two-thirds of the money with you and send you a check every month.

Now almost anybody who had a website had an easy way to earn some dinero.

If you had 100 visitors to your site last month, and only two of them clicked on an ad, you may have earned a grand total of 65 cents. If you got lots of traffic, however, the income could be substantial.

One spammer created a software program called Traffic Equalizer, which grew very popular. Traffic Equalizer was a "blender." It would grab sentences from websites all over the web and randomly mix them together on a web page with links or Google ads at the top. This program and others like it also randomly swapped out nouns, verbs, and adjectives so that Google would not recognize it as having been copied from somewhere else.

Traffic Equalizer pages didn't look like real English to a real person; they looked like gibberish. But search engine bots weren't smart enough to tell the difference back then. A guy with some blender software could churn out dozens of gigantic websites in a few hours. Pure, unadulterated spam.

Google and other search engines would then pick up these sites and list them. Now when you searched for something like "bad credit repair," a spam-generated site would come up alongside the legitimate sites. You'd land on a page that looked like this:

The most devilish thing of all was, since the page was all garbage, the fastest, most natural thing for the visitor to do was click on one of the links or ads, further increasing the spammer's profits.

I knew a guy who hired cheap labor in Ukraine to crank out hundreds of giant spam websites each month. He was collecting $70,000 per month in commissions, generated by ads on his spam pages.

It worked beautifully until Google and the other search engines caught on and figured out how to detect spammy pages and random-word robot blogs. Soon Traffic Equalizer stopped working entirely.

The Traffic Equalizer "content blender" programs didn't generate blogs that were as good as real human blogs, any more than hash busters generated emails that were as good as or superior to human-generated emails.

This illustrates a vital property of codes and language. The phrase "Been looking news make earth behind miles were no for bad credit report repair" is made from 100 percent real words, and some of it is even grammatically correct. But random word substitutions still give you gibberish. Random words are better than random letters, but they're still nonsense.

Random Letters Versus Random Words Versus Random Sentences

The email spam theory of evolution, then, is that random letter substitutions drive evolution. The blog spam theory of evolution is random words and sentences—which in the cell are random stretches of DNA transposed or transferred between cells.

Both theories assume natural selection and time are somehow capable, sooner or later, of gleaning real content from spam. Likewise, the "junk DNA" theory assumes 97 percent of our DNA is spam as well (628). The coding portion was assumed to be the small fraction that wasn't gibberish.

Based on our experience with search engines, do you and I have any reason to believe this is true?

Is there any such thing as a spam program that generates authentic English text that fools not only search engines but humans, too?

I've never seen a good one—though I've witnessed many attempts. Consider how Google ads evolve. In Google advertising, you improve your performance by split-testing ads. Here's a real-life example:

Popular Ethernet Terms
3 Page Guide - Free PDF Download
Complex Words - Simple Definitions
www.bb-elec.com

Response ("Click Thru Rate"): 0.1 percent

Popular Ethernet Terms
Complex Words - Simple Definitions
3 Page Guide - Free PDF Download
www.bb-elec.com

Response: 3.6 percent

Notice that in this ad the 36-fold improvement in response was achieved by simply transposing the second and third lines. That 36-fold is very, very significant because if the good ad costs you 20 cents per click, the bad ad would cost you $7.20 per click.

Here's another example from a real client's ad campaign:

Simple Self Defense
For Ordinary People
Easy Personal Protection Training
www.tftgroup.com

Response: 0.8 percent

Simple Self Defense
For Ordinary People
Fast Personal Protection Training
www.tftgroup.com

Response: 1.3 percent

The only difference between the first ad and the second is one word: "Fast" instead of "Easy."

Top-performing Google advertisers test hundreds, even thousands of ads. Fascinating as it is, it's a painstaking process. Is there any way for computers to write these ads, and save us humans all that time? What if you had a computer program that evolved your ads automatically?

It would seem that one should be able to write a computer program that swaps synonyms in for adjectives, like "Easy" for "Fast" or "Affordable" or "Convenient." It certainly would be easy to write a computer program that swaps lines 2 and 3. But that's exactly what the spammers have already done. And we all know what it looks like:

Spam.

To work effectively, these ads have to be crafted by a human, not by random computer substitutions. It may seem like a clever enough computer program could write ads as well as a human can... but if that program exists, I've never seen it.

What Can Natural Selection Do—and Not Do?

My friend and colleague Howard Jacobson, coauthor of the book *Google AdWords for Dummies*, performed a series of experiments where he demonstrates the evolution of a Google campaign, ad by ad.

Pay attention to the "CTR"—the click-through rate, Google's measure of the fitness of the ad:

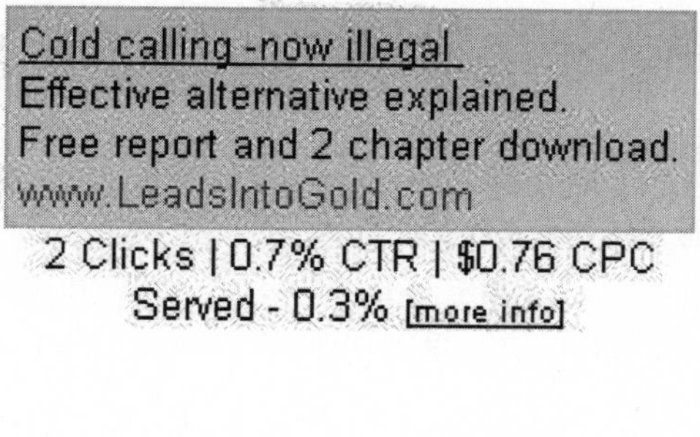

Cold calling -now illegal
Effective alternative explained.
Free report and 2 chapter download.
www.LeadsIntoGold.com
2 Clicks | 0.7% CTR | $0.76 CPC
Served - 0.3% [more info]

End cold calling forever
Small business marketing system.
Free report and 2 chapter download.
www.leadsintogold.com
430 Clicks | 1.7% CTR | $0.77 CPC
Served - 23.6% [more info]

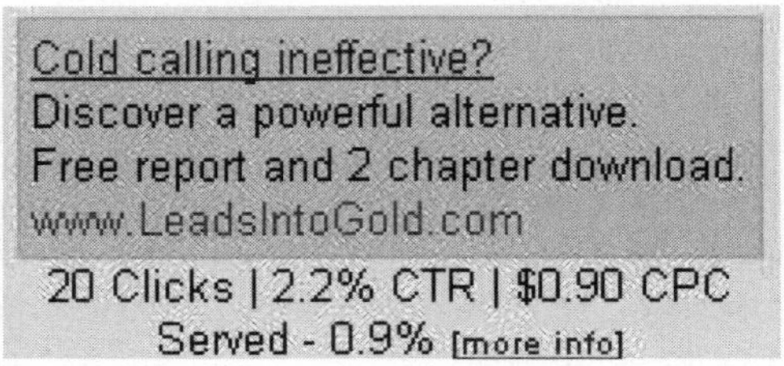

Cold calling ineffective?
Discover a powerful alternative.
Free report and 2 chapter download.
www.LeadsIntoGold.com
20 Clicks | 2.2% CTR | $0.90 CPC
Served - 0.9% [more info]

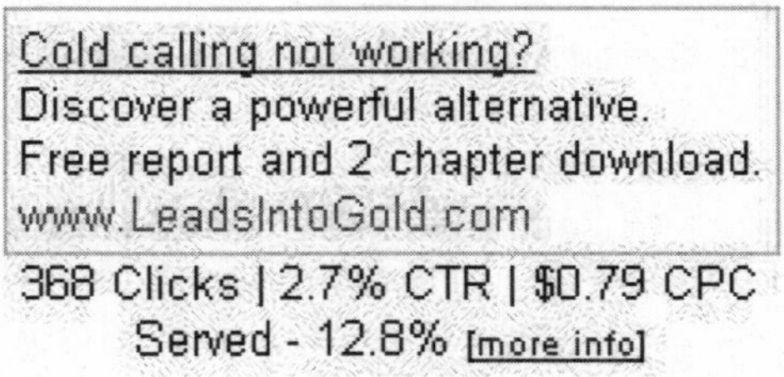

Cold calling not working?
Discover a powerful alternative.
Free report and 2 chapter download.
www.LeadsIntoGold.com
368 Clicks | 2.7% CTR | $0.79 CPC
Served - 12.8% [more info]

From beginning to end you see a 3.8-fold improvement in Click Thru Rate, from 0.7 percent to 2.7 percent. This means the winning ad gets 3.8 times more traffic for the same amount of money as the old. Click for click, the new ad will eventually cost 77 percent less than the original ad.

Google advertising is a Darwinian game where you write and rewrite ads. People click on some and not others. Google moves the winners to the top of the page, shows them more often, and sends the losers to the scrap yard.

When it comes to Google advertising, natural selection doesn't *create* anything. The people who write the ads do. The failure of one ad to get clicks, all by itself, does not in any way, shape, or form furnish fresh content for a different ad. In over 10 years of teaching AdWords to hundreds of thousands of people, I've also never seen anyone beat their old ad with text from a computer glitch or data copying error.

Natural selection is surely very powerful, and in advertising, it's the "court of last resort." Advertising pioneer Claude Hopkins coined that phrase in 1918, describing how the marketplace votes for or against your ads in maddening, frustrating ways that you never could have predicted.

But in the end, natural selection doesn't add, it subtracts. It kills off the losers. It's death, baby. That's all that it is. There ain't no life in death. That's why the phrase "evolution through natural selection" kills curiosity. It buries the most fascinating parts of the story—the fact that the cell, in carrying out Symbiogenesis, Transposition, and so forth somehow "knows" how to do its own genetic engineering.

Do you notice how many other things, like architecture, jazz, and video games, evolve purposefully, through the intention of their creators... not the way standard Darwinian theory has been telling us for decades? Ads never evolve by accidental copying errors. Neither do software programs, automobiles, music, or architecture.

Where does that creative input come from? In Google ads, the creativity comes from you, and the demands you hear from the marketplace.

As it turns out, the creative input in Evolution 2.0 comes from the cell.

The protozoan edits its own DNA much the same way you edit your ads: by monitoring hundreds of signals in its environment and responding with precision.

Charles Darwin was right about a lot of things... but not this. He thought cells were blobs of goo. He had no idea that tucked inside were fantastic networks of sensors, digital code, signal processing, and 24/7/365 adaptation.

Returning to our conversation about spam software... Spam software can't make great web pages. It can't write Google ads, either. Spam, even when it's generated from real English words, is still noise. It's still information entropy, and it's garbage. In other words, it's not enough to simply acknowledge that Transposition, Horizontal Transfer,

ONLINE SUPPLEMENT

How I apply the Swiss Army Knife to marketing problems in consulting projects

www.cosmicfingerprints.com/supplement

and the other blades on the Swiss Army Knife follow some interesting patterns. We also have to acknowledge that the system has *goals*.

If spam software can't write English intelligibly, how can cells give rise to even smarter, more powerful cells?

Are cells lucky spammers? Does natural selection somehow manage to turn Dalg, Glidj, and Blimlimlim into real words? Or are cells more like curious little monkeys that have the ability to read and write and predict and learn?

Answer's in the next chapter.

CHAPTER 19

Fluent in "Cellese"—DNA Is a Language

He was dancing to some music
No one else had ever heard
He'd speak in unknown languages
She would translate every word

—HARRY CHAPIN

FROM PREVIOUS SECTIONS, we know evolution happens. The question is *how*. So far, the answer is that cells engineer their DNA and form symbiotic relationships with each other.

We do not know *how* they know how to do this. We only see that they fight antibiotics, transpose genes, and form new species from hybrids every day.

Since you witness every day how human ideas evolve, you already have some notion of how cells *might* evolve. Could cells be like human ideas and evolve purposefully and intentionally?

When a bacterium is threatened by an antibiotic, does it switch on its internal genetic engineering systems because it's trying to stay alive? Does the conflict make it open to receiving useful sequences from other cells?

It may be that this is exactly what is going on.

Just the implications of DNA being code raise hackles for some people. But DNA is more than just a simple code. The pattern in DNA is a complex, intricate, multilayered *language*. An incredibly efficient and capable language at that. Not only is it a language, in much the same way that English and Chinese are languages, it appears to be even *more* sophisticated than human languages.

We also know that cells communicate with each other in other ways besides exchanging DNA. Since cells edit DNA, we know they "speak" its genetic language. This chapter sketches some of the features of this genetic language and offers references you can explore if you want to find out more.

Yes, I know—calling DNA a language is a bold statement. One person said to me, "Saying that DNA is a code or a language, etc., is a misinterpretation. Scientists and biologists use these words in an analogous way to explain genetics to non-scientific people." To him, I was making what he considered absurd comparisons between genetics and human language.

However, biologists and linguists alike have been comparing genetics with human language for decades. An entire school of thought in biology called biosemiotics (406) considers language to be a primary lens through which living things must be understood.

The 1984 Nobel laureate in Physiology or Medicine, Niels K. Jerne, used celebrated linguist Noam Chomsky's linguistic model to describe the human immune system (402). Another linguist, Gerald McMenamin, devotes an entire section in his book *Forensic Linguistics* to the language of DNA (515).

One reviewer of this manuscript, who teaches college linguistics, loudly objected to this. She said, "Biosemiotics is a joke. I'm sorry, but on this one, the facts just aren't there. [Speaking] as a linguist, not a biologist, cells do not use language. They may 'communicate,' but the requirements for language are not satisfied by what cells do. I teach this stuff.

"It's sort of like how people like to say dogs have a language or that pet owners come up with a language with their animals when it is not a language, as the criteria for complexity and abstraction are not met. Cells do not have the ability to use abstraction, therefore, it is very plainly not a language."

She was saying that bacteria send chemical "signals" but she was insisting it's a mistake to confuse these signals with symbols and language.

But research indicates that cells communicate in far more sophisticated ways than any animal we know of, in three distinct ways:

1. Internally, it speaks the language of DNA, and we know this because it extensively edits and rearranges its own genome;
2. Cells exchange linguistically coded chemical signals between members of their own species; and
3. Cells exchange linguistic messages with members of other species.

We know DNA expression is multilayered, and not merely chemical but abstract and symbolic (404). For example, AAA is not lysine; it's a *symbolic instruction* (326) to add lysine to the growing polypeptide chain—but *only* in the context of coding for proteins. Only about 3 percent of the genome codes for proteins. In most other contexts ("noncoding DNA"), AAA has different meanings.

Bioinformatics specialists are not misunderstanding linguistics; rather, the college linguistics instructor was underestimating the cell. This is a very common mistake.

Cells Speak "Cellese"

Rutgers University professor Sungchul Ji's excellent paper "The Linguistics of DNA: Words, Sentences, Grammar, Phonetics, and Semantics" (403) starts off, "Biologic systems and processes cannot be fully accounted for in terms of the principles and laws of physics and chemistry alone, but they require in addition the principles of semiotics—the science of symbols and signs, including linguistics."

Ji identifies 13 characteristics of human language. DNA shares 10 of them. Cells edit DNA. They also communicate with each other and literally speak a language he called "cellese," described as "a self-organizing system of molecules, some of which encode, act as signs for, or trigger, gene-directed cell processes."

This comparison between cell language and human language is not a loosey-goosey analogy; it's formal and literal. Human language and cell language both employ multilayered symbols (405).

Dr. Ji explains this similarity in his paper: "Bacterial chemical conversations also include assignment of contextual meaning to words and sentences (semantic) and conduction of dialogue (pragmatic)—the fundamental aspects of linguistic communication." This is true of genetic material. Signals between cells do this as well.*

Physicist and information theorist Hubert Yockey, writing in *Information Theory, Evolution, and the Origin of Life* in 2005, took great pains to demonstrate that terms like *translation* and *code* are literal, not metaphoric: "Information, transcription, translation, code, redundancy, synonymous, messenger, editing, and proofreading are all appropriate terms in biology. They take their meaning from information theory [320] and are not synonyms, metaphors, or analogies" (326). He says this on page 6, because this fact is fundamental to everything in his book that follows.

Human engineers build elaborate models and make exacting decisions in our designs. How could the cell be any different? How could evolution be any different? If evolution is true, cells must have some mechanism that knows how to rearrange their own "Russian dolls" in correct sequence. It's not a question of *whether* evolution happens. It's a question of *how*.

This would mean evolution occurs when rearrangements of DNA are organized based on a coordinated overall plan ("top down"), instead of by haphazard DNA changes in individuals ("bottom up").

---

* As an indication of the relevance of linguistics to genetics, go to http://scholar.google.com/ and search for "DNA genome linguistic." At the time of this writing, Google returns more than 28,000 books and papers. Introducing its book Introduction to *Biosemiotics: The New Biological Synthesis*, edited by Barbieri and Marcello, the publisher states, "Combining research approaches from biology, philosophy and linguistics, the field of biosemiotics proposes that animals, plants and single cells all engage in semiosis—the conversion of objective signals into conventional signs" (www.springer.com/us/book/9781402048135). A table in Ji (403) compares human language to cellese, showing both have alphabet, lexicon, sentences, grammar, phonetics, semantics, first articulation, and second articulation. The following is a simplified version of that table:

English	DNA Equivalent
Letters	4 nucleotides and 20 amino acids
Words	Structural genes
Strings of Words	Groups of genes coordinated by spatial temporal genes
Grammar	Folding patterns of DNA according to nucleotide sequences
Phonetics	Gene expression through protein binding and Epigenetics
Semantics	Cell processes driven by conformons and intracellular dissipative structures
Forming Sentences from Words	Sequences of gene expression in space and time
Forming Words from Letters	Organization of nucleotides into genes

The rearrangements would have to follow the syntax of the language. Evolution would have to be steered by the innermost Russian doll.

Does the majestic power of nature flow from the cell's ability to form and rework the logic encoded in its DNA? Is evolution possible because the cell *intentionally* adapts to its environment?

Francis Crick and Jim Watson uncovered the mystery of DNA in 1953. They figured out how the code is stored. But they only solved half the problem because physics and chemistry only supply the raw materials—they don't tell you where the instructions come from (326)!

Do cells read their DNA? Instead of "unthinkingly" carrying out their instructions, do they understand what it says? Might they have the ability to locate advantageous instructions from other cells' DNA and add those instructions to their own genome?

If so, then organisms would evolve their bodies by evolving their DNA language first. How might they "know" how to do this? Currently our understanding of this is very limited. However, we are beginning to clearly see what cells do.

As you've already seen, cells have the ability to rearrange their DNA and switch genes on and off, because of an aspect of the code that's different from the triplet code.

While it isn't clear how predictive or intentional this activity is, it seems cells are playing a complex numbers game with great skill. If it's language based, then surely it must be very similar to what humans do when we write poetry and music and design skyscrapers.

If a cell can choose to build a new linguistic pattern, then evolution itself is solved. And now the mystery of *why* and *how* it makes the choice is staring us in the face. Problem as of yet unsolved.

CHAPTER 20

Irreducible Complexity Made Simple

Remember the good old 1980s
When things were so uncomplicated.
I wish I could go back there again
And everything could be the same.

—ELO

DON'T YOU HATE IT when a busted five-cent part renders your entire Christmas present useless?

My friend Elizabeth's son Mark got a See 'n Say, the talking toy. That's the one where you pull the string and when the arrow points to the duck, it says, "QUACK!" He was enthralled with it until he tugged too hard and the plastic ring ripped off. The string zipped into the little hole, never to come out again. Without that plastic ring, the toy was useless.

The other day, the same thing happened to me when I lost the cap to our air mattress. Minus that cap, the mattress holds no air. My perfectly good, almost-brand-new mattress is now flat.

That's because of *irreducible complexity*.

Irreducibly complex systems are ones that can't function unless they have all their parts, assembled in the correct order. An air mattress isn't an air mattress without a cap.

Irreducible complexity is an argument often used by Intelligent Design advocates to disprove Darwinian evolution.

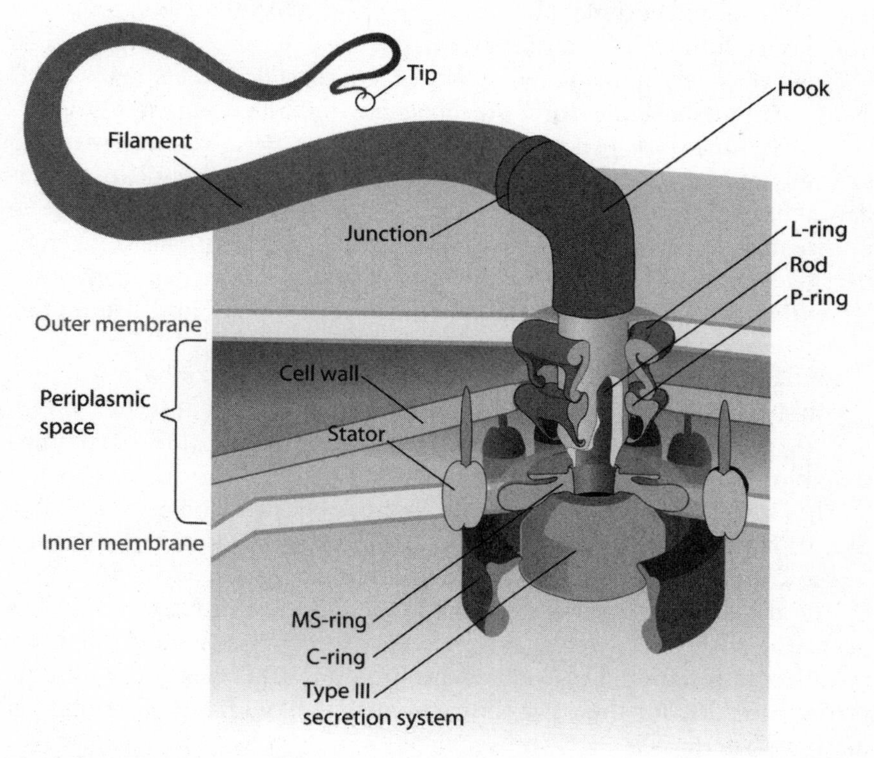

The bacterial flagellum is built from dozens of interdependent precision parts. The absence of any of them would render the entire mechanism useless. Such a structure can only evolve in steps if the sequences that code for the parts are just as modular as the flagellum itself.

In his famous book *Darwin's Black Box*, Michel Behe made this argument in relation to the bacterial flagellum. The flagellum is a 10,000 to 100,000 RPM propeller that bacteria use to swim. Behe insisted that the flagellum is useless if any one of its constituent parts is removed; if the hook, or stator, or filament, or any number of parts of the motor are missing, it simply doesn't work. Therefore it could not have arisen through the "numerous, successive, slight modifications" of Darwinism.

The flagellum is the poster child for Intelligent Design—it suggests that evolution by tiny gradual steps cannot be true, because half a bacterial flagellum (or even most of it) is only so many useless spare parts.

Numerous papers (e.g., 631) have challenged this, however, showing that the same genes that build these parts are present in other organisms, performing various other functions, and that subassemblies found in the flagellum are also found elsewhere.

However, *none of the papers that challenge the irreducible complexity argument about the flagellum solve the problem within the traditional gradual-mutation framework*. The papers that counter Behe do not seem to assume, let alone show, that the genes in question were built one accidental base-pair copying error at a time. They also generally manage to not mention that in order to assemble these parts from existing genes, those genes would have to be rearranged intact. These papers demonstrate that cells, just like words in English, Lego blocks, and automotive parts, are *modular*.

Predatory bacteria that pump toxins into their hosts have a motor assembly in those pumps that contains many of the exact same components you find in the flagellum (517). The same is true of the structures that allow cells to export or acquire DNA (510).

This shouldn't surprise you at all by now, because you already know that cells rearrange code with Transposition and Horizontal Transfer. These are all modular. They shuffle discrete blocks of code.

The flagellum is a perfect example of an adaptive invention that has been used over and over again for a variety of related transport tasks in different functional contexts. The bacteria didn't need to gradually evolve the parts for their flagellum because they could quickly make it using instructions already in existence.

While the flagellum may not have arisen, as Darwinism suggests, through those numerous small modifications, it could have arisen in the same way that a bacterium drowning in antibiotics assembles a pump on the spot: by finding and importing code from another bacterium that already had one.

Based on everything we've seen about Transposition, Horizontal Transfer, Epigenetics, and Symbiogenesis, it seems quite reasonable to hypothesize that bacterial ancestors built these subassemblies using their cognitive (662) and linguistic abilities. Eventually an exceptionally capable cell brought those subassemblies together to form the flagellum we know today.

Remember the original Evolution 1.0 formula?

RANDOM MUTATION + NATURAL SELECTION + TIME = EVOLUTION

Revise "random mutation" and the *real* formula becomes:

ADAPTIVE VARIATION + NATURAL SELECTION + TIME = EVOLUTION 2.0

Biology is full of systems that are irreducibly complex. Skeptics of Darwinism have been pointing them out for 100 years (904). But why should irreducible complexity be a problem for cells that are known to rearrange tens of thousands of blocks of code in real time *while the program is still running*? Evolution 2.0 is not a game of randomness and luck, but of blazingly brilliant software engineering.

There's so much more to variation than mutations; for example, symbiotic mergers and hybrids. Cells even steal code from viruses (652)!

Smart cells built the bacterial flagellum. Yes, the flagellum is irreducibly complex. And yes, Intelligent Design's opponents are right—it evolved from simpler, preexisting modular components, which were built from even simpler, modular components. But that flagellum wasn't built in a million lucky steps as traditional Darwinism claimed. Judging by the examples in the last few chapters, it was built in a very few generations by a very smart cell.

That bacterium wasn't lucky; it was *successful*.

BULLET POINT SUMMARY:

- Neo-Darwinism says Random Mutation + Natural Selection + Time = Evolution.
- Random Mutation is noise. Noise destroys.
- Cells rearrange DNA according to precise rules (Transposition).
- Cells exchange DNA with other cells (Horizontal Gene Transfer).
- Cells communicate with each other and edit their own genomes with incredibly sophisticated language.
- Cells switch code on and off for themselves and their progeny (Epigenetics).
- Cells merge and cooperate (Symbiogenesis).
- Species 1 + Species 2 = New Species (Hybridization). We know organisms rapidly adapt because scientists produce new species in the lab every day.
- #Evolution in 140 characters or less: Genes switch on, switch off, rearrange, and exchange. Hybrids double; viruses hijack; cells merge; winners emerge.
- Adaptive Mutation + Natural Selection + Time = Evolution 2.0

PART V

ORIGIN OF INFORMATION: THE QUEST

CHAPTER 21

Origin of Life: Can I Get a Straight Answer from Anybody?

The buzzard told the monkey,
"You are chokin' me
Release your hold and I will set you free."
The monkey looked the buzzard right dead in the eye and said,
"Your story's so touching but it sounds just like a lie."

—NAT KING COLE

NOW THAT INFORMATION SCIENCE has disproved the randomness theory of Evolution 1.0—and we have reviewed Evolution 2.0's highly organized Swiss Army Knife that engineers new adaptations—we're faced with an urgent question: *If randomness didn't create life, what did?*

In 2005, the legendary atheist and evolutionary theorist Richard Dawkins debated Design advocate George Gilder. The debate was on radio station WBUR in Boston (101). It was produced for National

Public Radio's *On Point* program. The show streamed on the web and I tuned in.

Dawkins at the time was a professor at Oxford University. One of his admirers had created a special endowment for him, The Charles Simyoni Chair for the Public Understanding of Science.

One of the callers asked Dawkins about the Origin of Life. He replied that it was "a happy chemical accident."

A *happy chemical accident*?

What kind of answer was that? And this is Oxford's "Professor of the Public Understanding of Science"?!

What if Isaac Newton had watched the apple fall out of the tree, and instead of formulating a theory of gravity, he had proclaimed it a happy accident? I was shocked Dawkins didn't get laughed right out of the studio.

"You're Overlooking the Freaking Obvious. *DNA* is the World's One Natural Code!"

A good buddy of mine said, "Perry, your challenge is easy to beat. I just say 'DNA' is the answer, and leave it to *you* to prove otherwise. I don't need to support my end—it's *obvious* that life is naturally occurring."

Not so fast, bro. Nobody gets to just *assume* they know the answer to Earth's grandest mystery, simply because they don't feel like shouldering the burden of proof. That's an unexamined worldview. A cop-out. It's also "begging the question": assuming the very thing you're supposed to prove... then offering it as proof.

Nobody knows how you get life from nonlife. Nobody knows how you get codes from chemicals. So nobody gets to *assume* life just happens because you have some warm soup and a few billion years; we want to understand *how and why*. Similarly, we can't just *assume* there's life on other planets; you have to prove that, too. Real science is based on inference from repeatable experiments. Anything less is abdication.

Huge scientific discoveries and quantum technological leaps lie in the question of Origin of Life/Origin of Information.

So What Do We Know About the Origin of Life?

I needed to find out. I bought a stack of Origin of Life books.*

Developing a theory for the Origin of Life is incredibly tricky because it must not only explain how the cell itself could come into existence, but also how it could replicate itself from the word "go." Anyone who tries to make it sound easy is not being straight with you.

In this book I do not even consider the question of what kind of process it might have taken to physically build even a "simple" cell with enough parts to function. All cells have hundreds of thousands of moving parts. I have attempted only to consider the question of how the instructions for the daughter cell got into the DNA, and how the language for those instructions was formed. Answers to even that simple, obvious question proved elusive.

We know that evolution requires cells to make copies of themselves. Before the daughter cell can be built, the genetic code must be copied. Before code can be copied, it has to exist. Evolution requires code. Code makes evolution possible.

Where Does Code Come From?

We must ask this question, because no Theory of Evolution can explain where code comes from. It only raises bigger questions:

---

* Books I read included Cairns-Smith (207), Küppers (224), Loewenstein (512), Bird (206), Hazen (219), and Yockey (326). Werner Loewenstein's book *The Touchstone of Life: Molecular Information, Cell Communication, and the Foundations of Life* was among the better ones. A great deal of the book is devoted to information systems in biology. Much of it is quite good.

However, on page 33, Loewenstein describes where the first information came from:

> The genesis of molecules, though sped up, still proceeded at a stately pace—that serenity one day would be shattered by an unprecedented event: among the mutants, one turned up—one in a myriad—who was able to catalyze the synthesis of his own matrix. The product here donated the information to construct its template! So, information flowed in circles: from matrix to product to matrix—*an information loop where the product promoted its own production.*

It's not obvious to the casual reader that the scene he describes is entirely theoretical, with *lots* of missing details. Is it even possible for such a thing to happen?

"Didn't DNA Evolve from Much Simpler Self-Replicating Molecules?"

Many people who prefer not to invoke a designer have asked this question. It's a valid hypothesis. But it's no more than that. Outside of DNA, there is no other self-replicating molecule known to science.

I did find some reports that claimed to be about self-replicating molecules (212). At best, though, these molecules proliferate through a process resembling crystal growth. *None properly self-replicate.* A big issue here is: Where do the subassemblies required for DNA replication come from? Most theorists put replication before protein synthesis, but you can't have one without the other.

How Do Most Origin of Life Researchers Answer This Question?

The most popular Origin of Life theory is the RNA world hypothesis. There are many kinds of RNA; the best known is messenger RNA, which we discussed in chapter 7. Messenger RNA is a molecule that's transcribed from the original DNA strand, and the message it contains is read to produce amino acids.

The RNA world hypothesis proposes that self-replicating RNA molecules (ribonucleic acid, a simpler, single-stranded molecule that resembles DNA) were the ancestors of current life (216). RNA strands store genetic information like DNA. According to this hypothesis, RNA was eventually superseded by the DNA, RNA, and protein world we have now.

Some scientists favor the RNA hypothesis because it seeks to explain the building blocks of life with as few components as possible.

Some viruses use RNA as their genetic material, rather than DNA (241). DNA is believed by some to have become the preferred mechanism for data storage due to its increased stability. This is supported by the observation that many of the most critical components of cells, the ones that are most common and evolve the slowest, are made mostly or entirely through RNA.

There are problems with the RNA world hypothesis: (1) Many scientists believe RNA is too complex to have arisen without the presence of the very same life forms it is believed to have created; (2) RNA is inherently unstable, so even if it did arise, it wouldn't last long without a cell to protect it; (3) catalysis of chemical reactions is seldom observed to occur in long RNA sequences only; and (4) the catalytic abilities of RNA are limited (204).

The RNA world hypothesis doesn't actually solve the chicken-and-egg problem of RNA and proteins: You need RNA to produce proteins, but you need proteins to build the machinery to read the RNA in the first place.

As a communication engineer, my objection to the RNA hypothesis is that to evolve any kind of cell, RNA would have to self-replicate. But the RNA strand formation you read about in the literature is not code-based self-replication. It's similar to crystal growth, which does not use codes at all!

RNA strand formation in a chemical lab is not in any way, shape, or form the same as DNA transcription and translation (252, 240). In DNA transcription and translation, in order to convert code to proteins, you need a ribosome to transcribe the message (137). But in order to have a ribosome you have to have a plan for building a ribosome first. A ribosome is partly made from RNA. So before that, you have to have a code *in* the RNA.

Many books and papers on the Origin of Life only discuss the assembly of the chemicals themselves. Nothing we know about chemicals tells us where codes come from.

Saying you can get real DNA by stringing chemicals together is like telling your kid that TVs come from a glass factory. The information content *in* the DNA is paramount. So in my research into current theories on the Origin of Life I focused on books and research papers with an information-centric view of life's beginnings. I found two kinds:

1. Some sounded as though scientists are just around the corner from some hugely promising, earth-shattering discovery.
2. The rest admitted that all current explanations for life's origin barely qualify as science.

To their credit, all gave better answers than "happy chemical accident." Most were still pretty disappointing, though. Many bulldozed right through the problem of information, making this assumption or that, such as this excerpt from a well-known, otherwise excellent biology textbook:

> In the laboratory, illuminated FeS_2 particles suspended in solution with an electron donor and carbon dioxide behave like photoelectric cells. When this or similar source of photo-chemical energy in membrane-bounded cells converted carbon dioxide and nitrogen to organic compounds, the biological revolution began. (231)

Keep in mind, the authors are describing one of the most glorious moments in the history of the universe. But they offer a terse non-explanation, then quickly brush on by as though the Origin of Life is a fairly minor problem, and a solution may present itself any day now.

Other people made statements like this—a gentleman posted this layperson's version of the RNA theory on my blog:

> Amino acids form spontaneously under a variety of conditions... Once we have replication (and mutation via faults in replication), we have evolution. Some RNAs go on to produce proteins and DNA, which turn out to be advantageous...

"Spontaneously"... "go on to produce"... "mutation via faults in replication"... "We have evolution!" is the RNA hypothesis in a nutshell, but it contains no mention of code at all.

The preceding explanation might sound okay until you remember that a digital code must be established before any kind of self-replication can be possible (see chapter 8). A code will only function in the context of an encoder and decoder. Plus, amino acids aren't code. A string of nucleotides all by itself is not a code.

Chemicals all by themselves don't communicate. No one has ever demonstrated that chemical reactions alone can generate codes. It's not nearly enough to have "hardware." You have to have software, too. Remember, the genetic code is crucial to all life and its ability to reproduce.

Without code there can be no self-replication (252).

Without self-replication you can't have reproduction.

Without reproduction you can't have evolution or natural selection (108).

Decades-Long Quest for the Origin of Life

The rest of the books I found openly admitted the Origin of Life is a complete mystery. Most of them would go on to say, "We may never know what happened in the earliest days of life, but . . ." and then go on to speculate.

It grew apparent that Origin of Life research had made shockingly little progress since the discovery of DNA. After billions of research dollars, it might be the most unsuccessful pursuit in all of biology. Biologist and complexity theorist Stuart Kauffman summed it up nicely: "Anyone who tells you that he or she knows how life started on the earth some 3.45 billion years ago is a fool or a knave" (223). And in 2011 John Horgan wrote a piece for *Scientific American* called "Pssst! Don't Tell the Creationists, but Scientists Don't Have a Clue How Life Began" (222).

Lynn Margulis said, "To go from a bacterium to people is less of a step than to go from a mixture of amino acids to that bacterium" (638). One is hard-pressed to find a single Origin of Life theory that can even be properly *tested* in the laboratory, never mind one that's been *proven*. I couldn't locate a single materialistic theory that explains the Origin of Information.

The only source that offered a rigorous approach to Origin of Information was Hubert Yockey's book *Information Theory, Evolution, and the Origin of Life*. It was a breath of fresh air. Yockey has been one of the world's most prolific specialists in bioinformatics and information theory, and I liked his book because he never flinches at important questions. Yockey includes a chapter that carefully considers how the present genetic code might have evolved from a simpler code. Whenever he speculates, he always makes this clear.

Yockey shows that the rules of any communication system are not derivable from the laws of physics. In his words, the origin of life is *possible* but not *knowable*. Yockey means that there is nothing in the laws of physics that prevent codes; they just don't explain how you get them. It's the same as saying the laws of physics are sufficient to allow your computer to operate, but they don't explain where it came from.

Yockey writes, "I have no doubt that if the historic process leading to the Origin of Life were knowable it would be a process of physics and chemistry," but "there is nothing in the physico-chemical world that remotely resembles reactions being determined by a sequence and codes between sequences." In other words, nothing in nonliving physics or chemistry obeys symbolic instructions. Yockey's conclusion is that, therefore, "the process of the Origin of Life is possible but unknowable."

He also says that there is nothing even resembling codes in water or rocks or chemistry. That's why there's not more in this book about chemistry. Chemistry doesn't answer these questions.

Yockey carefully explains every step of his reasoning and shows why our current framework of materialistic science, by definition, can never answer the Origin of Life question. However, he resists taking his findings to their logical conclusion. He never asks the question, "What *do* we know about where codes come from?" He disqualifies a designer from consideration at the outset, but the implications of his findings are clear enough: All that we observe about the creation of codes directly implies a designer.

Wanna Build a Cell? A DVD Player Might Be Easier

Imagine that you're building the world's first DVD player. What must you have before you can turn it on and watch a movie for the first time?

A DVD.

How do you get a DVD? You need a DVD recorder first.

How do you make a DVD recorder? First you have to define the language.

When Russell Kirsch (who we met in chapter 8) created the world's first digital image, he had to define a language for images first. Likewise you have to define the language that gets written on the DVD, then build hardware that speaks that language. Language must be defined first.

Our DVD recorder/player problem is an encoding-decoding problem, just like the information in DNA. You'll recall that communication, by definition, requires four things to exist:

1. A code
2. An encoder that obeys the rules of a code
3. A message that obeys the rules of the code
4. A decoder that obeys the rules of the code

These four things—language, transmitter of language, message, and receiver of language—*all* have to be precisely defined in advance before any form of communication can be possible at all (320).

But that's only part of the picture. Consider the whole system:

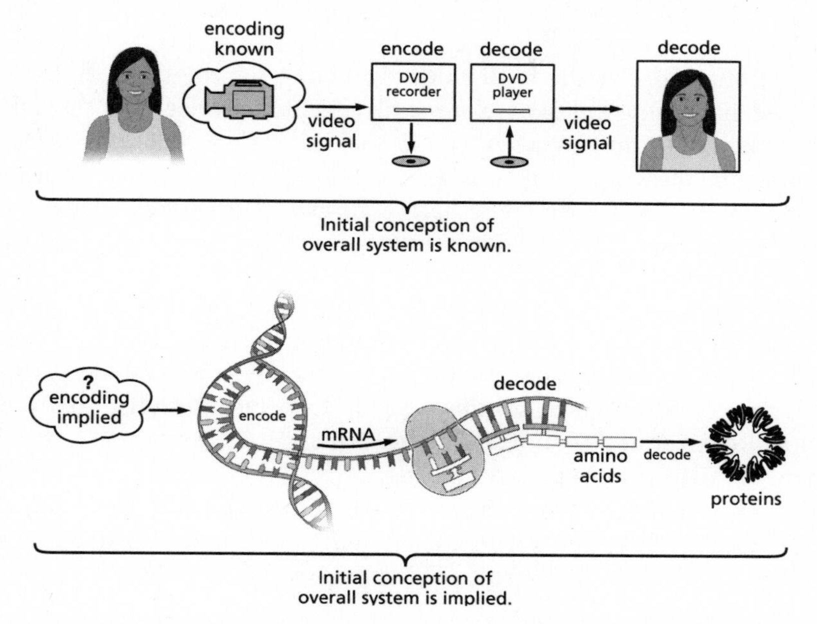

A camera sends a signal to a DVD recorder, which records a DVD. The DVD player reads the DVD and converts it to a TV signal. This is conceptually identical to DNA translation. The only difference is that we don't know how the original signal—the pattern in the first DNA strand—was encoded. The first DNA strand had to contain a plan to build something, and that plan had to get there somehow. An original encoder that translates the *idea* of an organism into *instructions to build* the organism (analogous to the camera) is directly implied.

The rules of any communication system are always defined in advance by a *process of deliberate choices*. There must be prearranged agreement between sender and receiver, otherwise communication is impossible.

If you leave out any of these things—code, encoder, message, decoder—your system doesn't work and can't evolve. There are no known exceptions to this.

By definition, a communication system cannot evolve from something simpler because evolution itself requires communication to exist first. You can't make copies of a message without the message, and you can't create a message without first having a language. And before that, you need *intent*.

A code is an abstract, immaterial, nonphysical set of rules. There is no physical law that says ink on a piece of paper formed in the shape T-R-E-E should correspond to that large leafy organism in your front yard.

You cannot derive the local rules of a code from the laws of physics, because hard physical laws necessarily exclude choice. On the other hand, the coder *decides* whether "1" means "on" or "off." She *decides* whether "0" means "off" or "on." Codes, by definition, are freely chosen.

The rules of the code come before all else. These rules of any language are chosen with a goal in mind: communication, which is always driven by intent.

That being said, conscious beings can evolve a simple code into a more complex code—if they can communicate in the first place. But even simple grunts and hand motions between two humans who share no language still require communication to occur. Pointing to a table and making a sound that means "table" still requires someone to recognize what your pointing finger means.

Which brings us to the question of the Origin of Life. How was the original signal, the pattern in that first DNA strand, encoded? What happened the moment the rules of language were first chosen?

Origin of Life Versus Evolution

Many scientists classify Origin of Life and evolution as entirely separate questions. Evolution we can study both directly and indirectly. But Origin of Life, bluntly, is still a freaking mystery. It's a field characterized by much speculation and precious little solid evidence. As you know from the beginning of this chapter, hardly anything we know about Origin of Life is empirical science. The only thing we know experimentally is that life comes from life.

One scientist asked why I move so seamlessly between these two questions.

I said, "Based on everything we *do* know about language, codes, systems, and engineering, evolution's mechanisms and Origin of Life cannot be separated. The reason why is that information entropy is your enemy. It doesn't matter whether you're *creating* codes or *evolving* them, natural processes are working against you. Whether you're creating codes or adapting them to new circumstances, you must create new rules. The only thing we know of that creates or re-creates codes is volitional beings. In all man-made systems, the only thing that makes them evolve is... intelligence."

I continued: "Randomness doesn't create codes, and once you do have a code, randomness can only destroy it. The quest for Origin of Life is nothing less than a search for a new law of physics. Ditto with the directional force that causes cells to evolve. Pretending that known laws do the job is an act of self-deception."

Information entropy is your ever-present foe. It raises the stakes even further, because the ability to evolve is necessary just for life to maintain.

Any life form without the ability to repair DNA copying errors would fast perish. And anything worthy of being called "life" must be able to not only replicate, but repair and adapt. No life form lacking the ability to evolve would last long (240). And we also know evolution is not automatic, not a "given," because everything we have experience with degrades and decays when left to its own. So life is very special.

The Origin of Information question is not merely an interesting academic question. It's the crowning scientific question of our time.

Someone asked me, "How do you know DNA couldn't be the one code that was formed by accident?" I replied, "Any of us can take that position if we wish. But it's not a scientific theory until science finds evidence of codes being formed by chance. There is no such evidence."

Codes are not matter and they're not energy. Codes don't come from matter, nor do they come from energy. Codes are information, and information is in a category by itself.

I had to find out where information comes from. Did anyone know? I was about to find out.

CHAPTER 22

Fistfight on the #1 Atheist Website in the World

This is why
Why we fight
Why we lie awake
And this is why
This is why we fight

—THE DECEMBERISTS

BY THE TIME I'd made my discoveries about information, DNA, and life's origins, my brother Bryan had become agnostic. He had little interest in "evangelizing" his views, to me or anyone else. As I began to unearth answers, Bryan grew less willing to spar with me. In the interest of preserving our relationship, he was reluctant to continue to engage these questions. Frankly, he wasn't all that qualified to have a deep scientific discussion anyway, since his schooling wasn't in science.

But I still needed a sparring partner. Deep down, what I really wanted was to be able to punch someone and get them to punch right back. Someone really smart. Someone relentless, who wouldn't indulge any of my nonsense, someone who would challenge every fact and assumption I've shared so far. Iron sharpens iron. I knew that I might come up with all kinds of views that could satisfy my own belief system. But we're all prone to self-deception.

I had come to an early conclusion that codes require designers. My scientific background spurred me to see if this would survive the scrutiny of others. I knew the internet could supply all the debate partners I could ever want. And even though most people obviously were not going to be the Ultimate Sparring Partner, there were surely a few out there somewhere. Needles in haystacks.

Without having motivated people to bounce ideas off of, I could not develop ideas about evolution and Origin of Life that would survive scrutiny.

Since I made a living consulting on Google AdWords, I used Google's advertising system to place ads on the internet. I took my intellectual pursuit to market, with ads such as this one:

> Origin of the Universe
> Did the Universe Come from God?
> Interpreting the Latest Results
> www.CosmicFingerprints.com

These ads appeared all across the web, especially on sites related to science and astronomy. I drove traffic to the Cosmic Fingerprints website as well as to my other site, CoffeeHouseTheology.com, where visitors could opt into email series with names like "Seven Great Lies of Organized Religion" and "Where Did the Universe Come From?"

In the first year alone, 30,000 people signed up for those emails. Every reply went straight to a dedicated email box. It seemed like everyone who got my emails felt an irresistible urge to argue with me about something.

A flood of website traffic opened thousands of conversations with challenging people—not just laypeople, but biologists, doctors, physicists, people of hugely diverse backgrounds. It was rather unusual. I was buying traffic to the tune of 1,000-plus visitors a day, instead of waiting for visitors to just show up. I was provoking people with a series of automated messages, then getting their replies.

Within a few years that list swelled to more than 150,000 subscribers. This put me in a very unique position of encountering a vast range of views. I was getting hundreds of emails every month from people of every conceivable belief system and opinion: Darwinists, Intelligent

Design advocates, Young Earth Creationists, atheists, Hindus, Jews, Muslims, Christians, New Agers, and mystics.

Everyone had an opinion about evolution. I was answering something like a hundred emails a week, and invariably, one or two of those exchanges would go deep.

I was sometimes outwitted by my opponents. These conversations shifted some of my views. For example, I had begun with a knee-jerk reaction *against* all living things having a common evolutionary ancestor, simply because of my Young Earth upbringing.

But then I read hundreds of conversations with people of all stripes, about things like whale feet and blind mole rats and pseudogenes and a hundred other things. Then, eventually, the remarkable discoveries of Barbara McClintock and Lynn Margulis. The evidence they offered slowly persuaded me that the case for some kind of evolution was credible.

I saw another kind of value in opening this up to the general public: Albert Einstein said, "It should be possible to explain the laws of physics to a barmaid" (905). This was also a place where I was learning to explain my own ideas to everyday folks.

My crazy experiment had the additional advantage of taking place in complete privacy. That was because one-to-one email is so personal. In the beginning, this was *not* taking place on a blog or a public discussion forum. Real names and email addresses, not screen names. Whenever someone wanted to converse, I insisted on honest dialogue and respect, not name calling.

Not everyone could abide by those rules! (It is impossible to have an honest dialogue with an anonymous person. This is why discussion boards almost always deteriorate into name-calling slam-fests. For that reason, on my blog I demand that people use their real names. This makes our discussions infinitely more civil.)

Our conversations explored a huge range of questions. One guy I conversed with was a columnist on one of the world's largest atheist websites, studying philosophy at a large university. Our conversation eventually mushroomed into a 120-page Microsoft Word document getting passed back and forth dozens of times as we responded to each other's challenges.

I spent many hours every week sifting through questions, responding and seeing if the positions I took could bear scrutiny. I figured, if anybody can overturn my discoveries, sooner or later that person is going to show up—it's just a matter of time!

I cannot possibly express how much I learned by doing this. I got a generous sampling of the beliefs of thousands of people all over the world, and obtained a deeper appreciation for the principle that *wherever possible, you should prefer a simple explanation over a complex one*—especially after witnessing the elegance of a crisp, one-paragraph idea versus four pages of ramblings!

For the first year or two, I chose to be neutral on evolution. Up to this point evolution had been a secondary question for me, taking a back seat to Origin of Life and Origin of Information.

A year into this journey, one of my friends, Andy Martin, heard about my Origin of Information research. He invited me to speak at Willow Creek, the largest church in Chicago. He was the organizer of Truth-Quest, their monthly forum for people who enjoy talking about tough questions, including evolution. In June 2005 I gave a talk at Willow called "If you can read this, I can prove God exists" and posted the MP3 and transcript at www.CosmicFingerprints.com/proof.*

That 2005 talk was the result of one solid year of presenting my ideas online. Hundreds of people had pounded the slag off my facts in private before I presented them in public. But I still dreaded opening my Cosmic Fingerprints email box, because I always knew I was in for a fight. My email conversations were neither easy nor fun. My learning curve was so steep, sometimes I got whiplash.

Atheists and Infidels

A few months after I posted my talk, a gentleman named Rob sent me an email. He was a fervent evangelical atheist. He had listened to my 2005 talk and came out swinging: "Perry, I see right through your sophistry and pseudoscience . . ."

We began an intense exchange. After a couple of weeks, he got flustered. One day in August 2005 Rob posted a link to my talk at Infidels.org—at that time, the world's largest online atheist community. (The Infidels forum was later taken over by a different website.) Rob basically

* In this talk, I use the word *prove* in the scientific sense (induction), not the mathematical sense (deduction).

said by way of introducing me to Infidels, "Be nice to this guy while you rip him to shreds."

I laid out my theory for the infidels to gorge upon:

1. DNA is not merely a molecule with a pattern; it is a code, a language, and an information storage mechanism.
2. All codes are created by a conscious mind; there is no natural process known to science that creates coded information.
3. Therefore DNA was designed by a mind.

I'd be lying to you if I said I wasn't nervous. My anxiety was off the charts. One of me, hordes of them. One slip of the foot and they'd eviscerate my sorry carcass like a pack of wolverines.

The anger and hostility was so thick you could cut it with a knife. Infidels was Grand Central Station for nonbelievers in God. These guys were *motivated*. I thought, *It's do-or-die time, Perry. If there's a hole in your theory, sooner or later these guys* will *find it*. What happened was fairly surprising . . .

Before this, I would never have imagined that a group of college-educated men and women would actually try to tell me that DNA isn't really a code.

But that's exactly what they did.

The atheists tried to tell me DNA was *not* a code. Then they tried to tell me a snowflake *is* a code! (If it is, what does it say?)

A lot of their arguments sounded sort of like this one from "Greyline":

> [Perry] is trying to pretend like DNA is code, and not chemical reactions. There's no need to bring in communication theory or the semantics of the word "code," or to make analogies with computer programming or languages or anything else. I do understand the usefulness of comparisons, but to base an entire argument on an analogy (like pmarshall's) serves no purpose. Just get back to basics—DNA is a chemical that self-replicates, imperfectly, and is therefore subject to natural selection. The result is what we call "life," in all its varied forms. Designer not required.

I spent an entire afternoon in the Oak Park, Illinois, library sifting through a stack of biology books two feet high. I cited textbook after textbook (313, 211, 215, 512, 311, 302) that explained how DNA is not just chemical reactions but is formally a code, and why it is a code based on universal definitions. I quoted Watson and Crick's acceptance speech at the Nobel Prize ceremony in 1962, where they were recognized for their discovery of the genetic *code* (303).

That trip to the library established the definitions I gave you in chapter 7. But no matter what I said, how detailed my explanations or how many references I cited, the Infidels would not accept it.

They lambasted me for taking the dictionaries and textbooks literally. Soon the moderator stepped in and began challenging me, too. I answered his every question; eventually he went silent and refused to respond. (A sidebar in chapter 7 gives a precise explanation for why the pattern in DNA is a code, and why the word *code* is not an analogy.) I limited myself to standard scientific reasoning and firmly established definitions from engineering and biology.

I wasn't ramming religion down their throat. I strictly limited my discussion to established science.

The end result of all this hot debate? After months, and eventually years, not a single hole punched in my research... even though this was an "open book test." Every opponent had ready access to all the books, websites, and scientific papers on the web—and still they could not counter with any evidence.

They steadfastly refused to accept any fact that might seem to support any kind of purpose in nature. After months of discussion, the Infidels, logic seemed to run something like this:

1. God does not exist.
2. Code implies God.
3. Therefore DNA is not code.

Really?!

It only took a few weeks for the atheists to run out of arguments and grow repetitious, but I spent two more years answering every single question and addressing every objection. I posted an exhaustive Q&A summary at http://www.cosmicfingerprints.com/dna-atheists/. You can click to six different pages that address

ONLINE SUPPLEMENT

How to handle militant opponents in hand-to-hand online combat

www.cosmicfingerprints.com/supplement

all the major points if you're so inclined. You can also click to the Archive.org snapshot of the forum.

Eventually I was thankful for my fistfight on the world's largest atheist website—even though I hated it at first. The debate tremendously clarified my definitions and strengthened my case. Without the Infidels, this book would not be possible.

All this further persuaded me that all we presently know about Origin of Life clearly infers a designer. I came to this conclusion because of the utter absence of any chemical explanation for Origin of Information. The logical inference was: **(1) The pattern in DNA is a code, (2) all the codes whose origin we know are designed, so (3) therefore we have every reason to believe DNA is designed.**

None of the thousands of people I interacted with offered hard evidence to support any other explanation. I posted these points on my blog, www.cosmicfingerprints.com. I bought even more clicks on Google and began to receive tens of thousands of visitors each month to the website. Arguments about my challenge exploded across hundreds of websites, blogs, and forums. The Infidels forum continued and eventually surpassed 100,000 page views. Doubters poured in from all over the globe and challenged me. The debates would go 'round in circles, sometimes for months.

Just like on Infidels, people still would insist snowflakes are codes; DNA isn't really a code; molecules are codes; sunlight is code. On and on it would go. Some people were not willing to accept Claude Shannon's basic definitions and have a discussion on common ground.

One day while conversing with an especially stubborn skeptic, I had a crazy idea: *Perry, why don't you offer him $10,000 if he can show you a naturally occurring code. Tell him precisely what he needs to give you. Give him 10 grand if he can deliver it.*

What happened next was even more interesting.

BULLET POINT SUMMARY:

- Neo-Darwinism says Random Mutation + Natural Selection + Time = Evolution.
- Random Mutation is noise. Noise destroys.
- Cells rearrange DNA according to precise rules (Transposition).
- Cells exchange DNA with other cells (Horizontal Gene Transfer).
- Cells communicate with each other and edit their own genomes with incredibly sophisticated language.
- Cells switch code on and off for themselves and their progeny (Epigenetics).
- Cells merge and cooperate (Symbiogenesis).
- Species 1 + Species 2 = New Species (Hybridization). We know organisms rapidly adapt because scientists produce new species in the lab every day.
- #Evolution in 140 characters or less: Genes switch on, switch off, rearrange, and exchange. Hybrids double; viruses hijack; cells merge; winners emerge.
- Adaptive Mutation + Natural Selection + Time = Evolution 2.0
- DNA is code. All codes whose origin we know are designed.

CHAPTER 23

Information: The Ten-Million-Dollar Question

There are more things in
Heaven and Earth, Horatio,
than are dreamt of in your philosophy.

—WILLIAM SHAKESPEARE

MY SKEPTIC was especially persistent. I was citing well-established information science and he was refusing to accept it. I was getting annoyed. Suddenly I decided, "I'm going to lay down the gauntlet. I'm going to offer this guy 10 grand if he can solve this."

To make an offer like that—an offer of financial compensation for the discovery of a naturally occurring code—my first step was to write a specification telling him exactly what he was supposed to find. *OK, dude. Here's exactly how you prove me wrong.*

I consulted a standard engineering textbook for definitions (321). I used the definition to show why a computer language is a code; I showed that DNA is also code for identical reasons; then I went on to outline what I was asking for—a code that met the same requirements, that was not designed. (I've included that specification for you in appendix 4.)

The document described how any person could demonstrate this, according to the criteria from the engineering book. I posted it at www.NaturalCode.org.

On the blog I challenged him and everyone else: "If you can find a naturally occurring code that meets the spec at naturalcode.org, I'll write you a check for $10,000.00."

He poked me with a few questions and then...

... disappeared.

The round-and-round in circles about "sunlight is a code" stopped, too. And the "Perry, you unfortunately don't understand what DNA does" and "Code is just a metaphor that scientists use to explain complex chemistry to ordinary people"—all of it stopped. Just like that.

As soon as they worked the diagram on the website form, anybody who thought sunlight was a code quickly realized why it wasn't. (Try it yourself if you're still not sure.) Anybody who thought DNA was merely *like* a code and not an actual code, needed only read the spec. They could see for themselves how neatly DNA transcription/translation matches up with Claude Shannon's communication system.

A few weeks later, another person came along and posted a lengthy comment challenging my statements. Just like the other guy, I offered him the 10 grand. He did the same thing: He protested for a bit, then vanished.

The inference to design in DNA's transcription/translation process could no longer be swept under the rug.

For five years I issued the $10,000 challenge, usually to people who were extremely passionate about this issue. But in five years I never got a single submission.

If you want to dismiss the idea of design in biology, you have to demonstrate where information comes from first.

Whether we have access to the designer was a question for the theologians; to an extent I set that question aside. The science question was: *How* was it designed? And just how intricate was the design?

Every living thing comes from information in its parents' DNA. The cell reads information from DNA, and the DNA has the instructions for building the next cell. Any theory of evolution is a theory of how information gets created.

I did offer that first $10,000 simply to score a debate point. I admit it! But others took my question more seriously. They saw value in making such a discovery; it would drive science forward. This isn't just an academic question.

By this time I was already starting to work on the book you're reading right now. I thought, *Hey, if I could assign a* million-dollar *prize to this instead of a ten-thousand-dollar prize, that might draw serious attention to this question.*

I could write a check for 10 grand. Writing a check for a million, though—not so easy. I started asking my friends for ideas on how to secure backers for a million-dollar prize.

One night I had dinner in Chinatown with my friend Johann. Johann was a programmer who was very well familiar with these questions. He instantly understood my natural code challenge, and acknowledged the odds of someone solving this weren't very great.

I complained to him that even though it wasn't very easy for someone to win the million dollars, investors weren't exactly beating my door down to offer up their life savings.

He said, "Perry, I know how you can make the million dollars profitable for the backer."

"How's that?"

"Offer your investors patent rights in exchange for the prize money."

"Patent rights? For what?"

"Artificial intelligence."

"I don't get it."

He leaned forward. "Perry, if someone figures out how to stir chemicals in a tank, and without cheating, gets those chemicals to generate codes and talk to each other, they've created intelligence. They have discovered AI—artificial intelligence, the kind they talk about in science fiction movies! Most people have never really thought about it, but that old 'warm pond where the first spark of life occurred' theory is really a theory about AI."

His eyes brightened. "Intel or Apple or some biotech firm would pay a *lot* of money for this! Offer a million dollars for the patent rights. Then once you've got the patent, go sell it to some company in Silicon Valley for even more. Remember, Perry, they've been trying to come up with real artificial intelligence for years and it's been a long, hard slog. This would be a *major* breakthrough."

Dang. He's right.

I realized this was much more than a posturing tactic in a philosophical debate, far more important than scoring ego points. Origin of Information is a fundamental scientific question that demands an answer. It goes to the heart of the difference between living and nonliving things. It addresses the AI problem.

Suddenly my little $10,000 challenge—which I didn't really think anybody could win anyway—became exponentially more interesting. If someone *did* figure out how to do this, how to meet the spec, how to make codes emerge from pure chemistry without cheating, they would have a bona fide communication device—the very first one apart from living things that was not designed by a human!

As the technology was further developed, who knows what it might do?

Maybe it would become self-aware. It could give computers the ability to make free choices, something that is currently not possible. Right now, all computers can do is flip switches as they obey rules. Computers can't create anything. But if someone came up with chemicals or materials that can create code—the sky was the limit.

Another friend said, "Perry, one million is too small. Your prize goal should be $10 million." So I decided, when the book came out, I would also go to the public and make the following offer, which I'm calling the Evolution 2.0 Prize:

One hundred thousand dollars, out of my pocket, goes to the first person who discovers a naturally occurring code through any process, patentable or not.

If it is patentable, the $100,000 is a deposit. We file for a patent, and once the patent is granted, pay the remainder prize out of backers' funds.

It is low risk for the backers, because I get soaked for $100K before they have to put up the rest of the money. Then we sell our new patent, helping the discoverer navigate the corporate minefield of negotiations and enforcing patents. The commercial potential would be big enough for the backers to want to invest to begin with.

A number of people I told about my plan mentioned the Ansari XPrize, the $10 million award for a privately built spacecraft that could make two flights in two weeks. Contestants invested millions of dollars developing spacecraft. A team financed by Microsoft founder Paul Allen won the prize in 2004. So I began shopping for backers who would put up the money in exchange for the patent rights.

Today I'm offering this to you, the reader of this book. Discover a naturally occurring code, one that matches the description at www.NaturalCode.org, and you'll qualify for $100,000.

If we can secure a patent for your discovery that's solid and defensible, you get the remainder of the prize. To see details of the prize and the current purse amount, check www.NaturalCode.org.

"How Can I Transform a Debate with Skeptics into Something Useful for Humanity?"

Knowing science can never prove God, I had a startling realization: To be intellectually honest as we explore the question of Origin of Life, it is *necessary* to play both sides of the naturalism-versus-design fence.

Do I believe that life is designed, either directly or indirectly, that it requires intentionality to exist and reflects some kind of higher purpose? Yes, I do.

Is it possible science will overturn that theory? Perhaps.

Will science continue to reveal amazing details about how it all works? Absolutely.

So I must respect science and remain ever curious.

The Evolution 2.0 Prize respects both science *and* theology because it acknowledges an unanswered question that demands an explanation. You can't sweep Origin of Life under the rug and call it a happy chemical accident. Nor can you insist that it's beyond the realm of human discovery, as some Creationists do.

Theology believes God made a world that is understandable and discernible—as does science.

If we respect both disciplines, we can transform this heady, philosophical debate into something truly practical and useful.

By offering this prize I give both sides their due respect. If the discovery is made, humanity benefits by getting a new technology. And we get a deep glimpse into the mysteries of life itself.

About the Prize

The Evolution 2.0 Prize focuses on the real issue of the Origin of Life question: not "Where did the chemicals come from?" but "Where did the information in DNA come from?" Instead of asking "What about the hardware?," it asks "What about the software?"

I'm offering $100,000 to the first person who submits a purely chemical process that produces codes. If the process is also patentable, the prize awards up to $9.9 million to acquire the majority of the patent rights. The prize caps at 10 million dollars, balance to be paid when the patent is granted.

All communication systems, including DNA, have three distinct, interconnected parts: encoder, message, and decoder. The prize goes to the first person who discovers a natural process that produces a complete communication system without having to specify (design) the encoding and decoding rules in advance. Such a process, if discovered, would revolutionize modern science.

So far as we know, codes and communication systems are only created by minds. If someone can locate an exception to this, they've made a landmark, once-per-century–level discovery. Such a discovery would garner recognition far beyond the Evolution 2.0 Prize.

To see detailed specifications of the prize or to submit an application, visit www.naturalcode.org or see appendix 4.

Why Origin of Information Is Important and Valuable

Discovery of a naturally occurring code would be more important than it may initially appear. Let me explain why. In short, the person who discovers this will have discovered a way to produce artificial intelligence (AI). (In chapter 26 I discuss the technological implications of such a discovery.)

In 1953, Stanley Miller and Harold Urey at the University of Chicago mixed chemicals in a vat and produced amino acids (235). Since then, their experiment has been heralded as a major clue to the origin of life. My complaint with their experiment is, while it did create rudimentary building blocks of life, it didn't create any form of coded information at all.

Miller and Urey didn't solve the problem I raise in chapter 21: You need a DVD player to read the DVD, but you also need a DVD recorder to make the DVD in the first place. This means the existence of DVDs requires a simultaneous plan for recorders, DVDs, and players.

Origin of Life shares this same fundamental problem: the existence of life simultaneously requires the rules of a code, with encoder, message, and decoder.*

What if Miller and Urey's experiment, or one like it, did create codes? What would it have to accomplish? It would have to satisfy the specification that I've written in appendix 4. The most up to date version is at www.naturalcode.org.

Let's say you poured some chemicals in a flask, at some particular concentration of compounds, at the right temperature and pressure, and so forth. Let's say those compounds generated, transmitted, and received a simple code.

Perhaps one end of the jar encoded the temperature into a digital message, and the compounds at the other end of the jar decoded that temperature in reverse and displayed it in some code that we could interpret. Imagine that this happened without anyone "cheating."

This would mean that chemicals alone, without minds or brains belonging to humans or other living things, would have done something even computers, on their own, cannot do: build a simple communication system from scratch. A computer cannot imagine. It can only obey the instructions it's been given. Because a new code would have to be defined in advance (as with our DVD player), this jar of chemicals would be assigning meaning to symbols. It would be making a creative, linguistic choice.

That would be a very simple instance of real artificial intelligence. But from there it would only be a matter of time before we figured out how to vastly increase its sophistication. Someday Siri on your iPhone might become a real being that knows you're there, instead of merely an app that is programmed to recognize voice patterns.

Forget about validating the materialistic paradigm of science. The discovery of the naturally occurring code that the Evolution 2.0 Prize would reward is worth a lot of money to industry, possibly hundreds of millions.

* The broader question of life itself is a chicken-and-egg one. As one person asked me, "Why would a code appear before there was anything to code? To use your analogy, why would a television show appear before the television itself was around?" Based on everything we know about communication systems, a reason to communicate always comes first. If the universe is purposeless, there is no "why," so there is no reason for life to begin in the first place. Therefore, codes and life infer not only a designer, but purpose in the universe.

Is It Even *Possible* to Get Information from Matter or Energy?

It is often said that if you supply all the right chemicals, replication will "kick in" and then the inevitable forward march of evolution will begin. The enduring legacy of the Miller-Urey experiment illustrates this.

Not only is this hypothesis statistically improbable, it may be fundamentally impossible. Here's why.

In my search for an explanation of where information comes from, I was rocked by this statement by MIT's legendary "Father of Cybernetics," Norbert Wiener:

> "Information is information, neither matter nor energy. No materialism which does not admit this can survive at the present day." (324)

If Wiener, the famous MIT professor and mathematician, was right, then there are three entities in the universe: matter, energy, and information. Matter and energy are interchangeable (i.e., $E = mc^2$) but information only comes from consciousness. So far as anyone knows, information is not an emergent property of matter.

What did Wiener mean when he said, "Information is information, neither matter nor energy?"

He did not mean that information itself is spiritual or that it contains some sort of mystical essence.

He meant that it adds a dimension of order to matter that does not come from the matter itself. He meant that you can have a stack of paper and a jar of ink, but dumping the jar of ink out on the paper does not a book make. He meant that the difference between *A Tale of Two Cities* by Charles Dickens and *The Stand* by Stephen King has nothing to do with the paper or the ink.

ONLINE SUPPLEMENT

Stem cell specialist Robert Lanza, MD, explains biocentrism—the principle of consciousness first, matter second

www.cosmicfingerprints.com/supplement

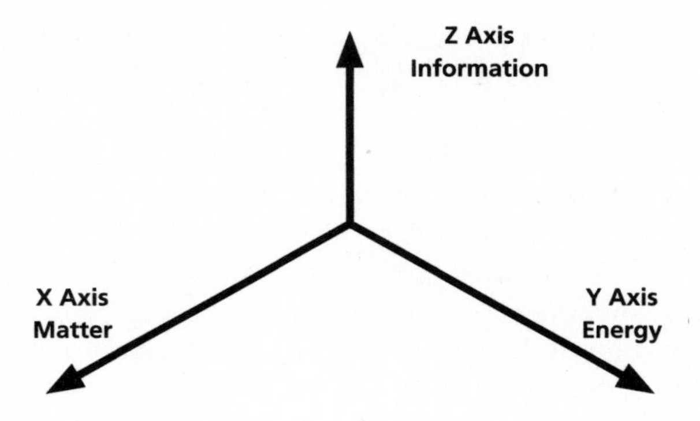

Matter, energy, and information are at right angles to one another. Mathematically, they are independent variables. We do know of ways to convert matter to energy and energy to matter, but currently we do not know any way to convert matter or energy into information without the action of a conscious being (324). The Evolution 2.0 Prize seeks to answer this question.

Before this, I had never noticed the distinct difference between matter and energy versus information. Suddenly I saw it everywhere I looked. I couldn't un-see it! If all you had was X and Y, there was no way to get Z.

Is a rock information? No.

Is a TV screen information? No.

Are the photons of light coming from the TV screen information? No.

Is the pattern formed by the light coming from the TV screen information? Yes.

How Do You Get from Chemicals to Information?

The existence of living things requires information (among many other things). As we discussed earlier, information systems are not an emergent property of matter or energy. They're a separate entity (323, 326).

All attempts to explain Origin of Life only in terms of chemicals trivialize the questions that most demand to be answered. The 1953 Miller-Urey experiment (235) produced organic compounds from gases thought to be present in Earth's early atmosphere. It is still widely cited in textbooks as an explanation of how early life was formed in the ocean.

The Miller-Urey experiment only attempted to explain where a handful of the chemicals came from, and it certainly didn't begin to explain how replication got started. How could it—without bringing in information and its critical role in living things?

A Shakespearean Example

A good way to explain this is with the question, "What is *Romeo and Juliet*?"

It's a play written by Shakespeare, of course.

Does *Romeo and Juliet* exist?

Of course. Romeo never existed and neither did Juliet, but the play about their love story certainly does.

But in what sense does it exist?

It's a code! It was encoded by William Shakespeare and appears in many forms: handwritten manuscript, printed book, Word document, radio program, string of 1's and 0's. *Romeo and Juliet* is *Romeo and Juliet* in all of those forms.

Shakespeare's play exists whether its words are stored on a hard drive or transmitted by radio waves, sound waves, or lasers in a fiber optic cable.

Romeo and Juliet is not paper. It's not ink. It's not sound or light. It is a *pattern* of words. The pattern itself is neither matter nor energy.

For an interesting thought experiment, try arguing that *Romeo and Juliet* is not real. Or that the money in your checking account isn't real. Or that the Japanese yen isn't real. Are all these things just figments of our imagination?

Information is a set of rules that operate *in addition to* the laws of physics. Wiener's observation means that codes perform real, measurable tasks and that, if the effect is real (e.g., your garage door opens) then the cause (the radio signal that tells it to open) must be real, too.

There's an infinite difference between information and non-information. There is no middle ground (326). It's either code or it's not. There's no such thing as a quasi-code.

Most atheists* embrace as their world view the theory of materialism—the idea that physical matter is the only reality and that everything,

* The home page of the Infidels website says, "As defined by Paul Draper, naturalism is 'the hypothesis that the natural world is a closed system, which means that nothing that is not a part of the natural world affects it.' Thus, 'naturalism implies that there are no supernatural entities'—including God." (Retrieved February 11, 2015)

including thought, feeling, mind, and will, can be explained in terms of matter and physical phenomena. Within this view, everything is believed to fit neatly into the laws of physics and no external causes or purposes, no prime movers, no gods or spirits are required.

But I could find no formula or transformation that turns matter or energy into information. This is precisely what the Evolution 2.0 Prize seeks to discover.

Freedom of Choice

Information possesses another very interesting property that distinguishes it from matter and energy. That property is freedom of choice.

In communication, your ability to choose whether "1 = on and 0 = off" or "1 = off and 0 = on" is the most elementary example of the human capacity to choose. Mechanical encoders and decoders can't make choices, but their very existence shows that the choice was made.

By definition, none of these decisions can be derived from the laws of physics because they are freely chosen. In the history of the computer industry, somewhere along the way, somebody got to *decide* that 1 = "on" and 0 = "off." Then everyone else *decided* to adopt that standard.

Physics and chemistry alone want us to be fat, lazy, and unproductive. Gravity pulls us down. Entropy makes us old and tired. Clocks wind down. Cars rust. Signals get static. LPs scratch. Desks become cluttered. Bedrooms get strewn with dirty clothes. Choice rises up against this.

Evolution 2.0, far from mindless, is literally mind over matter. The unfit adapt. Order and structure increase. Cells exert control over their environments. Underdogs come from behind and win.

Consider how information is measured. Distance is measured in meters, power is measured in watts, time in seconds, and mass in kilograms. But information is measured in bits. Eight bits = 2^8 = 256 combinations or possible *choices*. Each bit is the freedom to select a 1 or a 0. That's what makes it useful. Bits are choices!

Information capacity is capacity for choice. A choice is a totally different thing than a kilogram or a watt. That's why Wiener said, "Information is information, neither matter nor energy."

That means materialism cannot explain the origin of information, the nature of information, or the ability to create a code or language from scratch. It can't explain thought, feeling, mind, will, or communication.

The zinger is, 150 years post Darwin, it can't explain evolution, either.

So many mysteries remain unsolved. But that doesn't mean they're unsolvable. And though many people are content to invoke the hand of God and travel no further, I hope our sense of the divine will provoke *more* curiosity about science, not less.

I want our sense of the divine to drive us forward, to look for the answer to this question. It's not enough to throw up our hands and say, "God did it." We need to ask:

How?

That's the burning question that motivated the Evolution 2.0 Prize.

BULLET POINT SUMMARY:

- Neo-Darwinism says Random Mutation + Natural Selection + Time = Evolution.
- Random Mutation is noise. Noise destroys.
- Cells rearrange DNA according to precise rules (Transposition).
- Cells exchange DNA with other cells (Horizontal Gene Transfer).
- Cells communicate with each other and edit their own genomes with incredibly sophisticated language.
- Cells switch code on and off for themselves and their progeny (Epigenetics).
- Cells merge and cooperate (Symbiogenesis).
- Species 1 + Species 2 = New Species (Hybridization). We know organisms rapidly adapt because scientists produce new species in the lab every day.
- #Evolution in 140 characters or less: Genes switch on, switch off, rearrange, and exchange. Hybrids double; viruses hijack; cells merge; winners emerge.
- Adaptive Mutation + Natural Selection + Time = Evolution 2.0
- DNA is code. All codes whose origin we know are designed.
- Where do codes and linguistic rules of DNA come from? $10 million prize.

PART VI

EVOLUTION 2.0 AND ITS IMPLICATIONS FOR SCIENCE

CHAPTER 24

Beyond "God of the Gaps": A New Paradigm for Biology

Look out please, mind the gap
Watch out for the people trap
Here we are, going down
Hold on before we hit the ground
Look out please, mind the gap

—THE NOISETTES

STEPHEN HAWKING tells an important story in his book *God Created the Integers*. One of Isaac Newton's supporters asked the great scientist, "Could the solar system, with the planets all revolving around the sun in the same direction in almost the same plane, be formed out of an initial uniform distribution of matter by the action of only natural causes, or was it evidence of design?" (218).

Newton answered that his system could in no way explain these obvious regularities in the heavens, that they could not result from the action of only natural causes. The cause "had to be not blind and fortuitous, but very skilled in Mechanics and Geometry."

Stephen Hawking further relates:

> And so matters stood for nearly the entire eighteenth century until mathematician Pierre Simon de Laplace blazed his way

> across the firmament of French science. In 1770, Laplace began a rapid outpouring of papers on a wide variety of topics in pure and applied mathematics, drawing wide attention to himself. The most important papers focused on outstanding problems in planetary theory. The orbits of the two largest planets Jupiter and Saturn sometimes lagged behind and sometimes ran ahead of their predicted position. Laplace sought to explain how the planets influenced each other in their orbits. This is a more difficult problem than the three-body problem which even today can only be solved by successive approximations. Laplace demonstrated that perturbations were not cumulative, as Newton feared, but periodic. God did not need to intervene to keep the Solar System from collapsing. (218)

This is an important story because it demonstrates that, wherever possible, we should search for systematic processes, not consign them to the realm of the undiscoverable.

Hawking's story illustrates why many people have a problem with the theological perspective known as "God of the Gaps" (937)—in which gaps in scientific knowledge are taken to be evidence of divine intervention. That is to say, as soon as we don't know or understand something, then... "God did it."

It's a cop-out. We don't need theories that prevent scientists from doing their jobs. Upton Sinclair said, "It is difficult to get a man to understand something, when his salary depends on his not understanding it." When he said this, he was criticizing those who earned the salaries. But I can't fault scientists for resisting beliefs that undercut their profession. Money aside, scientists have an important job! We cannot afford to sideline critical questions.

"God of the Gaps" has an evil twin: "God Had Nothing to Do with It." This, too, is a cop-out. We've seen the anti-theist agenda in popular atheist authors who have grossly underreported the evolutionary process as they strive to purge goal-directed systems (teleology) from scientific thought. The result is that 99 percent of people have never heard an accurate "2.0" description of how evolution actually works.

The consequence is a century of people underestimating, underreporting, and even mocking the grandeur and power of nature. Dogmatic individuals decreed that evolution proceeds by random copying errors and declared 97 percent of our DNA to be junk. For decades, the establishment has mocked and punished anyone "naïve" enough to believe nature is purposeful.

We need to adopt a new pair of scientific guidelines:

1. Any theory that takes a job away from a scientist is probably wrong.*
2. Any theory that attempts to eliminate God as an ultimate explanation is probably wrong.

These two statements stand forever in tension with each other. The solution to the God of the Gaps problem is for us to always assume *ultimate* intentionality, logic, and order *without* assuming we've reached the end of the scientific rabbit hole. Nothing in science is "ultimate." No scientific discovery has ever turned out to be "the end of the road." The history of science shows us there's *always* another realm of order to discover.

While I am sympathetic to many critiques of the Modern Synthesis from the Intelligent Design movement, and while I have many friends who are ID advocates, ID too quickly jumps to "Designer" as an immediate explanation. The Designer is offered as a replacement for legitimate inquiry into detailed systems and fascinating natural processes.

ID as it is commonly understood threatens to take jobs away from scientists. ID pits faith against science. Until this changes, ID will *never* gain acceptance in the scientific community.

Anti-ID takes jobs away from scientists, too. Nobody's going to fund your DNA research project if they're convinced 97 percent of it is junk.

If, however, we take "Designer" to be an *ultimate* explanation, with an unknown number of layers in between, then both nature and God receive their due respect. Science is freed from the corset of reductionism and scientists gain greater reasons to pursue ambitious research

---

* This does assume that the theory in question is a legitimate line of inquiry in the first place. The public cannot afford to pay scientists to endlessly speculate. People stopped taking perpetual motion machines seriously even before the laws of thermodynamics told us why perpetual motion is impossible. That was because no perpetual motion machines were known to exist. While many have attempted to ban a designer from scientific discussions, design is still the only available inference for Origin of Information. The only way to find a natural explanation is to fully acknowledge the inference to design and thus outline the exact questions that a natural explanation must answer. This is why I feel it's more appropriate to offer a cash reward for Origin of Information *after* it's discovered, than to pour money into programs that have thus far failed to produce results.

programs. A Designer as an ultimate explanation *employs* scientists, instead of denigrating their work.

Some gap is always with us, and always has been. This is always uncomfortable, always a source of strife between fundamentalists on each side of the divide. It reaches back over centuries of scientific debates.

The gap we currently face is the one between living and nonliving things. Our newest scientific models have not successfully crossed this chasm. And the only hard evidence I have ever been able to find regarding this gap, as we've discussed, pointed to the Origin of Life being caused by an act of intelligence.

In other words, it's not a question of *whether* intelligence is involved; it's a question of *when* and *at what level.*

A New Master Paradigm for Biology?

Science is the formal practice of a grand hypothesis, a master paradigm. At the moment, that paradigm is as follows:

Current Master Paradigm of the Physical Sciences

1. Matter, energy, space, and time are organized according to discoverable laws of physics (933).
2. The laws of physics are unchanging and universal (910).
3. All past and future discoveries reveal law-like behavior exhibiting mathematical elegance and beauty.*

* "When I am working on a problem, I never think about beauty. I think only of how to solve the problem. But when I have finished, if the solution is not beautiful, I know it is wrong"; Richard Buckminster Fuller (quoted in 908). "Mathematics, rightly viewed, possesses not only truth, but supreme beauty"; Bertrand Russell (942). "The mathematical sciences particularly exhibit order, symmetry, and limitations; and these are the greatest forms of the beautiful"; Aristotle (916).

10 Presuppositions of Science

1. The existence of a theory-independent, external world
2. The orderly nature of the external world
3. The knowability of the external world
4. The existence of truth
5. The laws of logic
6. The reliability of our cognitive and sensory faculties to serve as truth gatherers and as a source of justified true beliefs in our intellectual environment
7. The adequacy of language to describe the world
8. The existence of values used in science (e.g., "Test theories fairly and report test results honestly")
9. The uniformity of nature and induction
10. The existence of numbers (933)

This current Master Paradigm of science (in physics) can never be absolutely proven in the ultimate sense, but it is certainly testable. That testing has been consistently rewarded. It has been extraordinarily successful in the study of nonliving things.

For 100 years, though, we've been trying to force biology to fit inside the physics Master Paradigm. It's not working. The Master Paradigm doesn't explain information. It doesn't explain why living things are *willful*.

Although the current Master Paradigm of science is a necessary *part* of the behavior of living things, it is *insufficient* to fully describe what they do. That's because of the yawning gap between nonliving and living things we just talked about.

It is imperative that we openly acknowledge the chasm between living and nonliving things. We can't pretend the chasm doesn't exist. If we do, we become like Procrustes, the thief and murderer from Greek mythology who would capture people and tie them to his iron bed, stretching them or hacking off their legs to make them fit.

We know of no way to make the physics of dead matter explain what living things do. It's time to stop stretching explanations that are nothing more than just-so stories. Amputating biology, trying to pretend that it is not purposeful and is really no different than studying rocks or snowflakes, is even worse.

I'm all for filling this gap; that's why I'm offering the Evolution 2.0 Prize. Meanwhile, in its absence, I want to highlight the urgent need for a new Master Paradigm—one that extends beyond the physical sciences and applies to life sciences (biology).

The eminent biologist Carl Woese, discussed in chapter 12 for his research in Horizontal Transfer and groundbreaking work in cellular evolution, lamented at how reductionist thinking has reduced biology to "become a science of lesser importance, for it had nothing fundamental to tell us about the world." He describes how biology has been shackled by the confines of reductionist physics, hoping that it will "press forward once more as a fundamental science." (680)

I believe not only that it will, but that it *must* if science is going to continue to advance.

In all my exploration I have never found evidence that life can come from nonlife. There is no known principle that bridges the chasm between the Master Paradigm of the physical sciences and the actual behaviors of living things.

It is time for Procrustes to stop chopping the legs off his guests whose legs hang over the edge of the bed—and buy them a new bed!

That's why I introduce:

A New Master Paradigm of the Life Sciences (Biology)

1. Living things always obey the universal laws of physics and chemistry. But these laws alone are insufficient to explain what living things do, because each living thing is a *self* and its behavior is intentional. (403, 637)
2. Life is based on codes, and the laws of codes are not fixed and universal; they are freely chosen and local (320, 326).
3. Each layer of any code infers a higher level of intent. Every future discovery will reveal teleological, code-guided behavior that exhibits linguistic and logical organization. (645, 675)

I am suggesting that biology must draw upon an additional level of scientific principles beyond traditional physics, including linguistics, information theory, and signal processing. Perhaps even art, music, and architecture.

I recognize this is audacious, maybe even outlandish. However, the difference between living and nonliving things is so profound that any 6-year-old can plainly tell the difference. Life is so radically different from nonlife, scientists struggle to even agree on life's definition.

Living things are willful. When they're hungry, they eat. When they're afraid, they flee. They are selfish. This is true all the way from cells to insects to elephants. A common example of selfish cellular behavior is when a body rejects organs after a transplant; the body correctly senses that the new organ is "not one of us." In autoimmune diseases such as lupus and rheumatoid arthritis, the immune system's ability to recognize "us" versus "them" has malfunctioned and the body attacks itself (652). That's why we can sensibly speak of "self" as not only applying to individual cells but entire organisms.

Biology is more than an information system, but information systems are more like living things than dead rocks. The New Master Paradigm isn't just true for biology, but any information system. No legitimate field of human knowledge should be excluded from our models of biology. And I extend this paradigm to information—acknowledging that in any information system, the laws of its codes are freely chosen, not fixed. Consider the following:

- A watch is built according to information in a plan (instructions) that existed before the watch.
- A tree is built according to the instructions in its DNA, which existed before the tree.
- In both cases, an idea precedes embodiment and the idea is represented, symbolized, and abstractly referred to by a code. The map is not the territory. GGG is not glycine. It's instructions to make glycine. In "noncoding" regions of DNA, GGG means something entirely different.
- Design is when an idea precedes its embodiment.
- Therefore, watches and trees are ultimately a product of design.

The similarity between DNA and computer codes is merely the *crudest* way to describe the difference between living and nonliving things. It's a difference we can all understand. The true differences are far more profound. Living things are so sublimely integrated and orchestrated, my crude comparisons barely hint at their superiority to human-made designs.

The problem with my Master Paradigm of Life Sciences (Biology) isn't that it's *too* different from the current Master Paradigm of Physical Sciences; it's that it's not different enough (675)!

Filling in the Gaps

Scientists justly exercise a philosophical commitment to the incredible explanatory power of science. They can't know that science will always uncover more truths, but their profession demands they assume it will.*

Science has an admirable track record of filling gaps. Scientists have a term, *methodological naturalism*, which says that in the lab, you don't assume miracles; you always assume natural processes. In science you don't get to resort to a *deus ex machina* and claim the hand of God was meddling with your experiments.

As long as we respect this, then science and God are not incompatible; God is no longer the cosmic science-stopper. Rather, belief in God reinforces order and rationality in science. God is our ground for the Master Paradigm of science—our hypothesis that the universe is ultimately orderly, discoverable, and rational.

Is Life a Miracle?

So far as we can possibly tell, based on every available inference we currently draw from computer science, information theory, and bioinformatics, the

---

* In other words, scientists have *faith* in the scientific method. To many, the word *faith* does not seem to fit here; they prefer *confidence* instead. That's because many people interpret faith to mean "blind faith," faith without evidence. Implicit in this objection is that people of religious faith believe what they believe with no evidence whatsoever.

Rare is the theologian who would agree with this. In my personal experience, religious faith is based on evidence, past experience, history, and reason. *Evidence* includes things like the observations in this book, the beauty of nature, the fine-tuning of the universe, archaeology, and history; *experience* includes things like documented healings, answered prayers, and extraordinary sequences of events far too remarkable to ascribe to chance; and *reason* includes things like moral and philosophical arguments.

My websites www.cosmicfingerprints.com and www.coffeehousetheology.com are both devoted to the evidence, experience, and reasoning of faith. No one should be expected believe in God without evidence, and I don't know why they should ever need to. My personal definition of faith is more akin to the confidence that a scientist expresses in science than any sort of blind commitment as some might assume. I explain why faith and science are necessarily interdependent at www.perrymarshall.com/godel.

origin of life has all indications of being a deliberate, astonishing act of outside intelligence—a miracle, you could say.

If a miracle by definition is an exception to the "regular and normal operation of things," then science by definition can't reduce a miracle to a process. No miracle will ever fit into science's neatly organized shelves and file folders.

If this is true, it might be impossible to test that theory in the lab.

Nevertheless we must all respect any scientists who have confidence that this gap will be bridged; they are only doing their job, after all. Thus, the motivation for the Evolution 2.0 Prize is all about bridging that gap. The person who wins the prize may have discovered the key to unifying physics and biology.

CHAPTER 25

Applied Evolution I: Which Teaches Us More? Darwinism Versus Evolution 2.0

I'm a jet fuel genius
I can solve the world's problems without even trying
I have dozens of friends and the fun never ends
That is, as long as I'm buyin'

—STYX

DARWIN'S DANGEROUS IDEA, as evangelized by Daniel Dennett, was just another perpetual motion machine. Evolution 1.0 was exciting in theory, unworkable in reality. Replication, variation, and selection are *not* enough to produce evolution. But if we ever *can* create systems that evolve the way cells do . . . watch out!

Do you recall imaginary numbers from high school math, where the square root of –1 is *i*? Such numbers were first believed to be an entirely theoretical construct of the mathematical world. But mathematicians found that they are incredibly useful in analyzing vibrating systems. Engineers who work with analog and digital circuits use them all the time.

Likewise, even highly theoretical concepts in physics have applications in engineering, like predicting what makes electrons flow from one part of a transistor to another. Theoretical constructs almost always find practical use sooner or later, *if* they're logically sound.

Jerry Bergman, former Bowling Green State University biology professor, asked: What practical skill do my students acquire by studying Neo-Darwinism (103)?

He had a hard time thinking of any.

If one believes that the entire principle of evolution is summed up in random changes and natural selection, then that directly implies that the only way to improve something is to make vast quantities of corrupt copies and let survival of the fittest sort them out. (Dear reader, let me say once again that most professional biologists have a far more nuanced understanding of evolution than this—but the general public hears little about it.)

But you know what really happens when you use accidental copying errors to write a software program, or if you make haphazard business decisions, or accidentally strike keys on a piano—natural selection will murder your endeavor very quickly!

Give it time—eons of time—says Darwinian gradualism, and eventually your musical score, business venture, or software will work out. But it might take a billion years. It's unfathomably wasteful.

Random mutation plus natural selection allegedly produces all kinds of fantastic designs, yet Darwinism remains useless to engineers. When students learn simplistic Darwinian evolution, they don't acquire a single practical skill...with one possible exception.

Genetic Algorithms

A genetic algorithm (GA) is a computer program inspired by evolutionary concepts. It simulates evolution on a computer. It modifies code and then evaluates the code against some preprogrammed goal, keeping the winners and discarding the losers. GAs refine software programs through an evolution-like process. GAs are not a be-all and end-all by any means; they have limited application, but they are useful.

Some years ago Dawkins wrote a famous GA software program to demonstrate how Darwinian evolution might successfully work (111). He entered the following random string of letters into the program.

WDLTMNLT DTJBKWIRZREZLMQCO P

One letter at a time, his program evolved this string of letters. After only 43 iterations, by randomly changing letters and deleting results it didn't want, the program reached its preprogrammed goal of the following sentence:

METHINKS IT IS LIKE A WEASEL

This was heralded as a success. However, Dawkins' software program was programmed to compare each new sentence to the *goal* sentence and either select it for continued "mutation" or reject it based on whether it more closely resembled the *goal* than the previous mutation. But his very own "1.0" Darwinian evolution explicitly forbids preprogrammed goals! So Dawkins' "Weasel" experiment had nothing to do with true Neo-Darwinism.

His program does vaguely resemble what cells do. But don't forget—Dawkins has always insisted that evolution is blind and purposeless. His program is anything but blind and purposeless; its goal is precisely defined from the beginning! What Dawkins actually proved with this experiment was: If you want to evolve, you have to start with a goal.

Stanford computer scientist John Koza—famous for his GA research—wrote a paper called "Genetic Programming: Biologically Inspired Computation That Creatively Solves Non-Trivial Problems" (706). He shows that a GA is capable of designing analog circuits, and even producing creative designs, which turned out to have been patented in the past by ingenious circuit designers. In other words, a GA could, from scratch, create something demonstrably new and valuable.

I wonder how many people noticed that Koza's GA first had to be given a very specific, narrow set of constraints and precise goals before it could work? In this regard, the experiment, like Dawkins', does not resemble old-school Darwinism. It's an alternative form of design.

From the standpoint of the naturalism versus design debate, all successful GAs "cheat" with preprogrammed parameters guiding the GA to a specific desired outcome. They *always* carefully protect the rest of the code from the same kind of random mutations that Darwinism claims cause evolution.

Every few days, apps in my smartphone update. For months, one of my favorite music programs would crash if I paused the song for more than a few minutes and then pressed Play. But just yesterday they released a new version and it updated automatically. Software bug gone. It's so easy to take this for granted! But having worked for and owned software companies, I know the hours, days, and weeks of disciplined testing that go into these updates.

Cells also edit bugs out of their software, but for over a century, Darwinists have taken evolution for granted as a "given" that just somehow happens by time and chance. Now we know the cell is more like an operating system that updates and rewrites its own software (608, 643, 663). I can hardly think of any research endeavor that would be more valuable to humanity than precision understanding of how the amazing cell works and adapts.

If Neo-Darwinism really is accurate science—if we really can take evolution for granted as an inevitable product of time and chance—then why don't they teach random mutation and natural selection in engineering school? You'll recall I asked that in the very beginning. If Dawkins were right, engineers everywhere could get a lot more done with a lot less pain and suffering!

How come famous GA programs like Tierra (709, 710) and AVIDA (700) are little more than academic curiosities? Why aren't they all the rage with engineers and programmers who have important jobs to do? If natural selection explains how everything came to be, then how come it doesn't teach you how to build anything?

Untold Story: Genetic Algorithms *Can* Really Work

The number-one problem with GAs is they get stuck. You can run GAs on 10,000 high-end computer servers for months, but like a chess game doomed to an endless cycle of check that never gets to checkmate, GAs converge to specific points and can't evolve beyond them (714). Picture a toy car that drives itself into a corner and rocks back and forth until the batteries die. That's what happens to GAs.

This is because, without preprogrammed goals, Darwinian GAs only accumulate mutations gradually. They're inherently unable to make quantum leaps. Contrast this with Barbara McClintock's maize plants, which immediately repaired their broken chromosomes on the fly, replacing damaged code with new code so they could reproduce.

Some GA efforts have gotten around this with modular programs—subroutines and separate processes. Richard Watson and Jordan Pollack (714) found that if they programmed cooperative Symbiogenesis mergers into their algorithm, it gained the ability to become "unstuck." Their program also incorporates a mechanism similar to Whole Genome Duplication (see chapter 16).

In fact, they cite a series of GA research projects over the last several decades. One was by John Holland, "the Father of Genetic Algorithms." He wrote a foundational book in 1975 called *Adaptation in Natural and Artificial Systems* (704). In it, Holland drew from mathematics, computer science, and systems theory and considers problem solving in engineering, business, and genetics.

If computer simulations of evolution have taught us anything, it's that gradual, accidental, "Darwinian" processes never succeed in "climbing Mount Improbable." But modular systems programmed to make sudden dramatic changes often do.

Most genetic algorithms are little more than academic curiosities. The ones that *are* functionally useful mimic evolution's Swiss Army Knife and Natural Genetic Engineering.

If you are beginning to enjoy thinking *outside* the narrow box of Darwinian dogma, there are many more practical applications.

CHAPTER 26

Applied Evolution II: Technologies from Better Evolution Research

I can't believe you're saying
These things just can't be true
Our world could use this beauty
Just think what we might do

—RUSH

THE WHOLE REASON accountants and engineers and social workers and criminal justice majors go to school is so that by the time they graduate, they'll have a predictive model of how the world works. Hopefully when they have big decisions to make, they will make decisions based on education and experience—not wild guesses.

We should expect no less of evolutionary models than we do of models in accountancy or engineering. An effective Evolution 2.0 theory should not only explain how living things evolve; it also could be directly applied to making human-engineered systems evolve—and maybe even to accounting and engineering. In science, technology, and medicine, we can glean an enormous amount of knowledge by studying nature's designs.* This study is called *biomimetics*, a growing field with its own journals and conferences.

The post-Darwinian evolution paradigm holds exciting possibilities for medicine, cancer research, and aging...as well as software, search engines, storage and transmission of data, and the very relationship between human and machine.

In this chapter, we'll consider how a more complete understanding of evolutionary mechanisms will revolutionize our society in the future. For example, some of the most daunting problems in medicine and biology become much easier to solve when they're viewed as software programming questions. Many of the hardest challenges in technology become much simpler when we recognize that nature has already solved them.

A Cure for Cancer

If you want to think of cancer as evolution run amok, you're not far off the mark. When a cell is corrupted (and becomes cancerous), it replicates until its offspring are consuming all the organism's resources.

Cancer cells rewrite their code and change their strategy on the fly. Cancer is the result of a malfunctioning evolutionary Swiss Army Knife.

Random mutations can cause cancer by destroying key information in the DNA program. Apoptosis (programmed cell death) fails to kick in. The rogue cell starts reproducing, but not according to the same plans as the rest of the body. It doesn't stop (227).

The errors that cause cancer are often self-amplifying, eventually compounding at an exponential rate, in something akin to a chain reaction. Minor errors become severe. Evolution's Swiss Army Knife grinds away, missing critical information that it needs, and the body can't rein it back in.

This rebellion-like scenario is an unfortunate case of "survival of the fittest" where the driving forces of evolution war against the body's design and enforcement of order. In fact, once cancer begins to develop,

* When presented with tricky problems in my private consulting work, many times the first question I ask myself is: "How does the Evolution 2.0 Swiss Army Knife already solve this problem in biology?" This reliably leads to productive and surprising answers. I include material on this online at www.cosmicfingerprints.com/supplement.

the cell's willful forces continue to drive the progression of cancer toward more invasive stages, called clonal evolution (234).

Cancer offers us a fearsome glimpse of the tremendous force of the evolutionary algorithm. Cancer is an evolutionary runaway train, and that train has powerful engines.*

When patients undergo cancer treatments, their doctors adjust the strategy according to how the cancer responds, because the disease seems to have a mind of its own. In a sense, it does. Cancer cells actively mutate to evade the threat. The cell's ability to cut, splice, and rearrange DNA becomes the cancer patient's own worst enemy.

In a war, the dumbest mistake you can make is to assume your opponent is stupid or incompetent, when in fact he or she is a brilliant tactician. Since the former Darwinian paradigm assumes that evolution is purposeless and not goal directed, it stops that cure for cancer from ever being found. We've been underestimating our adversary because cancer cells are not stupid. Far from it. They're equipped with tremendous innate adaptive powers. That's why they respond in such diverse ways to treatments.

Only as we fully appreciate the sophistication of cells' evolutionary systems will we move into a position to treat cancer effectively.

Longevity and Anti-Aging Research

Gail Tsukiyama's book *Dreaming Water* is the story of Cate, a recently widowed 62-year-old mother whose daughter Hana is 38 but has the appearance of an 80-year-old woman. Hana suffers from Werner's syndrome, a genetic defect that causes her to age at twice the normal rate.

In this novel, which faithfully portrays this horrible disease, mother and daughter struggle to make the most of each precious day together, knowing that time steals away at double speed.

---

* Interestingly, half of all cancer cells have a copying error in gene p53. This is the program that tells the cell to halt its reproduction cycle until all chromosomes have been copied correctly (239). When this program fails, mutant cells duplicate themselves and the copying never stops. Tumors grow out of control, choking and starving healthy cells. Cancer claims another victim.

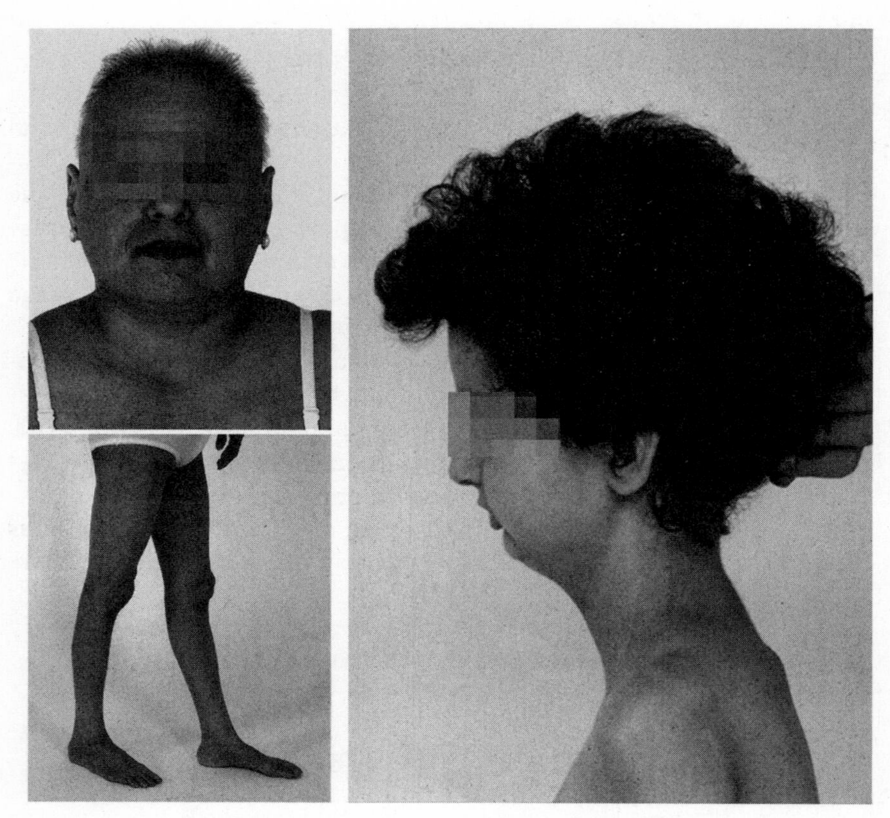

These women suffer from Werner's syndrome, a disease where cells are unable to repair damage to their DNA. It causes malformed body parts and accelerated aging.

Werner's syndrome is caused by the inability of cells to repair their own DNA. Cockayne's syndrome is similar, and also degenerates the nervous system. If we were born without the ability to repair DNA damage, we'd look 50 years old by the time we were 10.

If, on the other hand, our cells could always repair DNA damage, we might very well be able to live to be 500-plus years old. (Imagine the population explosion!) Aging is accelerated by loss of genetic information, the same information entropy we talked about in chapter 9.

If there is a "fountain of youth," it is to be found in understanding genetic repair systems and how and why they break down. As we learn to preserve these ingenious repair mechanisms, we'll look and feel younger and younger.

Artificial Intelligence

The movie *2001: A Space Odyssey* is an epic film about man versus intelligent machine. The spaceship *Discovery One* is piloted by a sentient computer, HAL 9000. The computer claims in a robotic voice that he is "foolproof and incapable of error," and later, when announcing a problem with the ship, insists the problem is "human error." The astronauts Frank and Dave go inside a vehicle where they can talk privately away from HAL, but HAL is reading their lips through a window.

While Frank attempts to repair the Earth radio transceiver during a spacewalk, HAL severs his oxygen hose, killing him. And when Dave returns to the ship with Frank's body, HAL refuses to let Dave in because he is jeopardizing their mission to visit Japetus, one of Saturn's moons.

Dave breaks into the ship and begins to dismantle HAL.

Spoiler alert: At first HAL apologizes and tries to reassure Dave. Then he pleads with him to stop, finally beginning to express fear, pleading in a monotone voice. Dave ignores him and unplugs each of HAL's memory modules one by one.

So…what would the true consequences of artificial intelligence (AI) be? If we succeed in creating machines with AI, will they be our friends or our enemies? If you want to explore the social implications of AI, science fiction authors will give you more than enough to think about. Meanwhile, we're faced with a tantalizing technical question: *How do you endow a machine with intelligence?*

Remember how cells talk to each other, engage in social relationships, and die for the greater good of the organism? Remember how they edit their DNA and form symbiotic relationships with each other?

The fastest path to understanding artificial intelligence (AI) is to understand the cell, because cells are cognitive (662). We can't know for certain now whether cells are self-aware or not, other than to say that some very competent researchers report that cells have impressive decision-making capabilities (518, 519, 509, 511).

Here's what we do know:

- Cells learn from past events.
- Cells grasp relationships between themselves, other cells, and objects in their environment.
- Cells exchange code with each other when triggered by certain conditions.

- Cells rewrite their own code.
- Some cells organize themselves into colonies. (523)

Thus far, nothing humans have ever made is as intelligent or willful as cells are. My computer is super fast and I love it. But at the end of the day, it's as dumb as a box of rocks. As the expression goes, "Garbage In, Garbage Out"—any computer is only as good as what you put into it. It possesses zero will of its own.

There is a very popular notion that someday, when computers get fast enough, they're going to "wake up" and become self-aware. But is there any evidence to suggest that this is actually true? Think about your own experience. Your computer is a million times faster than anything you might have had 30 years ago. But is your computer even 1 percent closer to being as "awake" as HAL 9000 was in *2001*?

On the other hand, the most fundamental unit of life, the tiny cell, already exhibits willful behavior. Surely, compared to studying the cell, all other paths toward developing AI are the long way around.

A biologist from the UK sent me this comment:

> Personally I think the evidence for Common Descent is strong and I certainly do not rule out that it could have occurred from a single cell. I think therefore that we both can agree that "evolution from a single cell" is possible.
>
> What seems undeniable is that there is evidence of intelligence somewhere in the process [of adaptation] and that the process itself seems to have a great deal of organisation.

I replied to him with a question: What if it's not just the intelligence of the cell, but the *networked intelligence* of billions of cells that makes adaptation possible? A network of cells is also exponentially more powerful than one by itself. We're only beginning to understand the systems of signals that travel between our cells.

Your brain exercises one kind of intelligence (decision-making ability and capacity to anticipate the future), while your immune system employs a different kind of intelligence. The intelligence that makes your body operate like a silent, well-oiled machine is not lesser than the intelligence of your brain; it's just different.

I'm pretty sure my body is smarter than my brain. If my brain were smarter than my body, I wouldn't eat so many cheeseburgers. And I would like to suggest that your body is not only smarter than your brain, but also every bit as intentional. It carries out functions with ease that we are seldom aware of and barely fathom.*

Technologists tend to be excessively optimistic. A few believe that all we need is enough speed and processing power and we'll achieve AI (707). Some believe that will happen spontaneously, when computers get fast enough; others believe it will take incredible ingenuity to make "strong AI" possible. But anyone who's interested in this subject will be well served to read what modern philosophers have said about it. The best book I've read on AI is *Brain, Mind, and Computers* by Dr. Stanley Jaki (705). It's especially impressive considering it was written in 1969. Jaki says intelligence is willful and intentional, and intent is an altogether different thing than computation. The question of AI asks whether mind and will have a metaphysical essence that is separate from their purely physical operation.

---

* Are cells smart or dumb? Is the human body smart or dumb? Is the body well designed or poorly designed? Skeptics often criticize the human body, presuming that it's an accumulation of chance accidents. They say things like, "The human eye is a pathetic design. It's got a big blind spot and the 'wires' are installed backward."

There are many, many variations on this argument. They're all variations on the "junk DNA" story.

When I was a manufacturing production manager, I had to produce an indicator lamp assembly for a piece of equipment. The design had a light bulb and two identical resistors, which I thought were stupid. I suggested that we replace the two resistors with one resistor of twice the value. This would save money and space. I told the customer, an engineer, I thought his design was poorly thought out.

The customer got angry and almost took his business elsewhere. My boss was livid.

What I didn't know was that 600 volts would arc across one resistor and cause a short circuit, but wouldn't arc across two. A second, "redundant" resistor was an elegant way to solve that problem and it only cost two cents. I learned the hard way that when you criticize a design, you may have a very incomplete picture of the many constraints the designer has to work within.

Designs always have delicate tradeoffs. Some have amazing performance but are extremely difficult to manufacture. Sometimes a minor change in material would make a huge improvement, but the material is unavailable. Sometimes you have to compromise between 15 competing priorities. It's a tricky maze to navigate.

I am not saying that there are no suboptimal designs in biology—I'm sure there are lots. Life followed an evolutionary process and many designs are "best guesses" engineered by the organism's ancestors. But human beings must be very careful to not proudly assert that we could "obviously do better." We don't know that. We do not understand what's involved in designing an eye because we've never built one. (Or, actually, we have, and they're all inferior.)

If you lose your eye, there's not a single scientist in the world who can build you a new one. Especially not arrogant speculators who try to tell you why the design of the eye is "pathetic." If I were selecting an eye surgeon, I'd look for one who has deep respect for the human eye, not disdain for it. How about you?

In my ignorance, I sneered at the customer's design, I found I could only appreciate it by approaching it with respect. A core value of biomimetics needs to be *nature is smarter than we are.*

What is this willful, self-preserving essence at the heart of living things? The Evolution 2.0 Prize strikes at the heart of this question. The possibility of an answer promises to alter the course of AI research.

Or, if there are no naturally occurring codes... strong AI may not be possible.

Materialism Versus Conscious Choice?

If materialism (the theory that our mind and will are determined entirely by their physical components) is true, then your ability to make choices is merely an illusion. What you believe is simply determined by chemical reactions in your brain, and you have no actual control over your behavior. If materialism is true, you might *appear* to be deciding what you think about this book, but you're really not deciding at all. Chemicals are making up your mind for you.

Do you say to your child, "Billy, I can't really blame you for sticking your hand in the cookie jar, because the chemicals in your stomach sent hunger signals to your brain"?

No civilization behaves as though our freedom to act is an illusion. For millennia, philosophers have understood that materialism offers no explanation for why it is possible for anyone to be a "free thinker" (922). More recently, in *The Grand Design,* Stephen Hawking says, "The molecular basis of biology shows that biological processes are governed by the laws of physics and chemistry and therefore are as determined as the orbits of the planets... so it seems that we are no more than biological machines and that free will is just an illusion" (924), and "Quantum physics might seem to undermine the idea that nature is governed by laws, but that is not the case. Instead it leads us to accept a new form of determinism: Given the state of a system at some time, the laws of nature determine the probabilities of various futures and pasts rather than determining the future and past with certainty" (924).

Hawking's book was widely criticized for shoving age-old philosophical questions under the bus in favor of his untested M-theory (911, 938). Hawking overlooks the fact that there is no known way to derive codes from chemicals, and that biology is more than just physics and chemistry. The Evolution 2.0 Prize focuses on this question. As for his view that we are only machines and freedom of choice is an illusion—to me this seems a disempowering view of the world.

So if materialism is incorrect, then we may never achieve true AI. But either way, we sure stand to learn a lot by trying to find out. This why it's so important to seek an in-depth understanding of the cognition that drives evolution. Barbara McClintock's question, "What does a cell know about itself?" isn't *Trivial Pursuit*. It's a lynchpin issue for science, philosophy, religion, technology, and medicine.

Presently we do not know whether our mind and will are made up of something else in addition to the components that allow for their physical operation. I embrace the "dualist" view, which asserts the human mind is nonphysical. I personally reject the materialistic view for three reasons:

1. I believe human beings really do make free choices and are responsible for those choices. This assumption is the foundation of legal systems and civilization.
2. You can't derive free choices from physical laws, because "free choice" and "physical law" are mutually exclusive.
3. Information itself is immaterial.

However, if we can discover a natural mechanism that creates codes—if codes and consciousness are *emergent properties* of simple physics and chemistry—then we can surely recreate that mechanism and design it into computers, cell phones, automobiles, and medical devices. AI would leap from science fiction to reality.

Test for Artificial Intelligence

It's hard for people to agree on precisely how to define intelligence, but British mathematician and code breaker Alan Turing invented a simple test for artificial intelligence: "A human judge engages in a natural language conversation with one human and one machine, each of which tries to appear human. All participants are placed in isolated locations. If the judge cannot reliably tell the machine from the human, the machine is said to have passed the test" (713).

Can you chat with a machine and be convinced the machine is actually a human? No machine to date has passed that test. The Loebner Prize holds a competition every year, offering a $100,000 award for the first "chatterbot" that judges cannot distinguish from a real human in a Turing test. This includes deciphering and understanding text, visual, and auditory input (http://loebner.net/Prizef/loebner-prize.html).

The prize has a five-minute time limit. Thus far, participants have only won the $3,000 bronze medal for making a respectable attempt.

No cigar for the big prize. Keep in mind that fooling a set of judges for five minutes is comparably easy; if you were to run such a test longer, the challenge would grow exponentially harder because the number of directions the conversation can go grows without bound. Imagine that you find a machine that seems pretty smart for five minutes... five hours later you're saying to yourself, "This is either a machine or a complete moron."

If AI is possible, not only would that machine be able to hold a conversation, it *might* also be able to write code. It would be a lot more like cells. It would surely have some ability to repair and rewrite code. A microchip that possessed artificial intelligence would quickly find its way into every conceivable consumer device. Millions of such chips would be harnessed to solve all manner of difficult mathematical, software, and database problems. All these chips could be connected to each other via the internet.

As soon as AI exists, computer programs will be able to evolve on their own by rewriting their software. The ability to self-evolve would be part of a good definition of AI, just as it's a function of a good cell. And any process that is claimed to solve the Origin of Life problem should be able to pass some version of the Turing test.

A machine that could pass the Turing test would also make a great search engine.

Intelligent Search Engines

Google's mission is to organize the world's information and make it universally accessible and useful. Billions of dollars of commerce hinge on how well this is done.

Organizing the world's information is an ambitious enough goal by itself, but the most challenging aspect of Google's job is to figure out what you really want based on what you type.

What if someone goes to Google and types in:

what if my girlfriend breaks up with me next week

Google doesn't "understand" this phrase. It is only endowed with a very sophisticated algorithm that matches the phrase you type in to current web pages. It ranks those pages according to their popularity and history of past visitors. It's all math. It identifies common phrases like

what If
my girlfriend
tries to
break up
with me
next week

and offers you some web pages with as many of those phrases as possible. If it can't find an exact match, it looks for phrases in close proximity to each other. It identifies which words historically signal more intention by the user than others.

When I typed that into Google, the first few entries looked like this:

Should I break up with my girlfriend? She gets mad at me for the...
What signs will tell me if my girlfriend is cheating...

Most of the entries are pretty reasonable. Still, Google struggles with anything that's not an exact match. What if the Google search engine actually, in some sense, understood what you were asking, the way a person does? What if it knew what you *meant* instead of merely matching *words that you used*?

That would be incredibly powerful. As good as Google is now, it would become 100 times more effective, even 1,000 or a million times more effective for some kinds of searches. For this one, Google would know the difference between him breaking up with her versus her breaking up with him, and only show sites about the latter.

The current approach to improving search engines is to build ever more sophisticated algorithms that add more linguistic rules to make more and more appropriate choices. But from an information theory point of view, that's a bottom-up approach. Genuine intelligence is top down; in other words, it begins with a strategy and a goal, organizing the smaller pieces to form a coherent whole, rather than simply logging correlations between search phrases and user behavior.

There's hardly a computer application where a higher form of intelligence wouldn't be tremendously useful. As amazing as Google is, a single cell possesses more intelligence than a multibillion-dollar search engine. The secrets are right under your nose, in every single cell in your body.

I suspect that if you solve the problem addressed in the Evolution 2.0 prize and patent it, Google would be happy to buy it for a considerable sum. Apple and Microsoft would kill to own a better search algorithm than Google's. So they would surely bid against them.

Self-Writing and Repairing Computer Programs

A spell checker can improve your writing, but it can't write for you. Why? Because all it does is apply rules. Rules dictate the structure of speech but they don't speak. Nobody can predict the next thing that's going to come out of your mouth by knowing the rules of English grammar.

My "show me a code that's not designed" challenge focuses on the question of how the specific linguistic rules of the genetic code came into existence. Equally important is the question of how the first genome came into existence. For cell replication, both are necessary.

Ever used Siri, the female voice-activated servant on an iPhone? Her uncanny ability to find the answer you're looking for is because of the patterns of millions of people who searched before you. Apple carefully monitors every search, paying attention to what happens next. Did you come back with a slightly different question five seconds later? Must have been a bad answer. Did you disappear for an hour? Must have been what you wanted, because apparently you were busy reading and enjoying that website. That's machine learning.

Language translation programs are the same way. They're based on millions of pages of translated text, and pattern recognition.

But what if that program could also create a new language? What if it could say something that had never been said before, simply because it *wanted* to? What if it could begin work on translating a new language, without ever being asked? What if Siri was a real *self* who understood her relation to you? Would your spouse be jealous?

Pattern recognition is one thing; program creation is quite another. Software companies still have to hire engineers to write the code, the program that recognizes those patterns. Your immune system doesn't. It writes its own code (661). The immune system judges which invading cells carry acceptable messages (antigens) and which don't, using high-speed pattern recognition. Wouldn't it be great if our software could do the same thing?

Personalized Technologies

What if your computer or cell phone were like a human personal programmer who does projects for you on command? What if you could say to your Android phone, "There's an iPhone app that's not available for Android, could you write a program that mimics it?" and it would come back to you two weeks later with a crude prototype. Perhaps it would inform you that it had collaborated with 1,500 phones that had gotten similar requests from their owners and were collectively evolving and rewriting the program for the new platform. Maybe, for instance, the phones would inform you they were developing a universal mapping program that would port all applications.

An entire industry and economy would spring up around this. Remember the 1999 dot-com bubble? The internet was taking off like wildfire and you had to be an idiot to not know this was a Very Very Big Deal. Everyone knew that somebody was going to get unimaginably wealthy from this; it was just a question of who.

Strong AI would spur another revolution as big as the internet (if not bigger), including investment bubbles and all the rest. But it would come with an additional ingredient that we've never had to consider before: The machines would want to get paid, too.*

To truly execute AI well requires more than just pattern recognition. It would require your phone to have an understanding of *itself* in relation to *you*. Intelligent phones would understand *self* in relation to other objects and goals, and thus act autonomously instead of only being acted upon by outside forces. What if your cell phone had the kind of intelligence that your dog has? Your dog knows what time you get home every day. Your dog senses your mood. Your dog has feelings. What if your personal technologies had something akin to feelings?

---

* Would they want money, or some other resource? Maybe they'd be like HAL 9000 trying to take over the spaceship. Maybe they'd be like the machines in the original *Matrix* movie and enslave the humans. Maybe they'll go on a quest for energy. Whatever the outcome, some clever science fiction writer will have anticipated it.

Yeah. I know that sounds creepy. Twenty years ago a lot of things we take for granted now would have sounded creepy, too—like any one of your hundred friends being able to land a vibrating text message right in your pocket anytime, day or night, or capturing a crime on video with your cell phone camera.

Ultra-High-Density Data Storage and Compression

Your cells contain at least 92 strands of DNA and 46 double-helical chromosomes. In total, they stretch 6 feet (1.8 meters) end to end. Every human DNA strand contains as much data as a CD. Every DNA strand in your body stretched end to end would reach from Earth to the sun and back 600 times (233).

When you scratch your arm, the dead skin cells that flake off contain more information than a warehouse of hard drives. Cells store data at millions of times more density than hard drives, 10^{21} bits per gram (242). Not only that, they use that data to store instructions vastly more effectively than human-made programs; consider that Windows takes 20 times as much space (bits) as your own genome. We don't quite know how to quantify the total information in DNA.

The genome is unfathomably more elegant, more sophisticated, and more efficient in its use of data than anything we have ever designed. Even with the breathtaking pace of Moore's Law—the principle that data density doubles every two years and its cost is cut in half—it's hard to estimate how many centuries it may take for human technology to catch up. Hopefully the lessons we learn from DNA can speed our efforts.

A single gene can be used a hundred times by different aspects of the genetic program, expressed in a hundred different ways (248). The same program provides unique instructions to the several hundred different types of cells in the human body; it dictates their relationships to each other in three-dimensional space to make organs, as well as in a fourth dimension, the timeline of growth and development. It knows, for instance, that boys' voices need to change when they're 13 and not when they're 3. It's far from clear how this information is stored and where it all resides. Confining our understanding of DNA data to computer models is itself a limiting paradigm. This is all the more reason why our standard for excellence ought to be the cell and not our own technology:

- DNA is a programming language, a database, a communications protocol, and a highly compressed storage device for reading and writing data—all at the same time.
- As a programming language it's more versatile than C, Visual Basic, or PHP.
- As a database it's denser than Oracle or MySQL.
- As a communications protocol it wastes far less space than TCP/IP and it's more robust than Ethernet.
- As a compression algorithm it's superior to WinZip or anything else we've dreamed of.
- As a storage medium it's a trillion times denser than a CD, and packs information into less space than any hard drive or memory chip currently made.
- And even the smallest bacterium is capable of employing all these mechanisms to dominate its environment and live in community with other cells.

Are you excited about this? Great; but you should be alarmed, too.

If you're paying attention, it should be plainly obvious that everything Evolution 2.0 promises to teach us can kill us even faster if we're not careful. Don't forget, HAL wants to kill Dave and take over the spaceship.

Only a view that nature is endowed with a higher purpose, and that we carry a responsibility of stewardship over nature, will lead us to limit our appetites and use technology responsibly. Most important, we need to sit at the feet of nature and let her teach us not only how to face the predators nature produces, but the ones *we* produce.

We can't run from the questions nature presents to us. We can't make up answers and call them science. We need to embrace nature in all her mystery. Then and only then will she reveal her secrets to us.

BULLET POINT SUMMARY:

- Neo-Darwinism says Random Mutation + Natural Selection + Time = Evolution.
- Random Mutation is noise. Noise destroys.
- Cells rearrange DNA according to precise rules (Transposition).
- Cells exchange DNA with other cells (Horizontal Gene Transfer).
- Cells communicate with each other and edit their own genomes with incredibly sophisticated language.
- Cells switch code on and off for themselves and their progeny (Epigenetics).
- Cells merge and cooperate (Symbiogenesis).
- Species 1 + Species 2 = New Species (Hybridization). We know organisms rapidly adapt because scientists produce new species in the lab every day.
- #Evolution in 140 characters or less: Genes switch on, switch off, rearrange, and exchange. Hybrids double; viruses hijack; cells merge; winners emerge.
- Adaptive Mutation + Natural Selection + Time = Evolution 2.0
- DNA is code. All codes whose origin we know are designed.
- Where do codes and linguistic rules of DNA come from? Evolution 2.0 prize.
- Answering this question will produce billion-dollar medical and technological breakthroughs.

PART VII

EVOLUTION 2.0 AND ITS IMPLICATIONS FOR FAITH

CHAPTER 27

The Real Reason People Don't Believe Evolution, and Why That Can Finally Change

It's a Barnum and Bailey world
Just as phony as it can be
But it wouldn't be make-believe
If you believed in me

—ELLA FITZGERALD

IN HIS BEST-SELLING BOOK *Why Evolution Is True*, the outspoken atheist and biologist Dr. Jerry Coyne relates this story:

> A few years ago a group of businessmen in a ritzy suburb of Chicago asked me to speak on the topic of evolution...After the talk, a member of the audience approached me and said, "I found your evidence for evolution very convincing—but I still don't believe it . . ." This statement encapsulates a deep and widespread ambiguity that many feel about evolutionary biology. The evidence is convincing, but they're not convinced.

Coyne bristles at this widespread phenomenon of disbelief in evolution:

> How can that be? Other areas of science aren't plagued by such problems. We don't doubt the existence of electrons or black holes, despite the fact that these phenomena are much further removed from everyday experience than evolution. (105)

If Coyne's business breakfast talk was anything like his book, then he made no mention whatsoever of Evolution 2.0's ingenious mechanisms, such as Transposition, Horizontal Transfer, and Epigenetics. Not a single word in 304 pages!* The only things "horizontal" are gorillas standing upright.

A Gallup poll (133) showed Coyne's observation about the general apathy and skepticism toward evolution to be accurate:

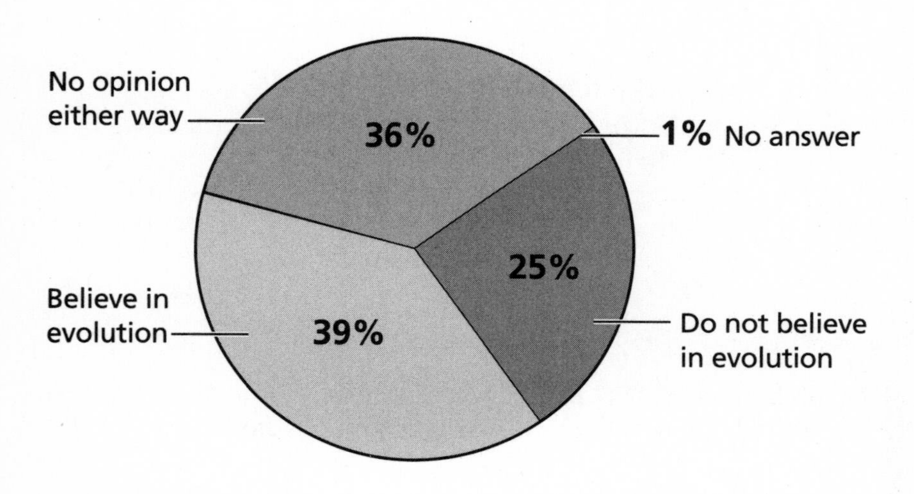

* He offers us a clue for why he made these omissions at http://whyevolutionistrue.wordpress.com/2012/08/22/james-shapiro-goes-after-natural-selection-again-twice-on-huffpo/ where he says, "Natural selection is the only game in town. Yes, we now know of a whole host of new mechanisms to generate genetic variation, including symbiosis and the ingestion of DNA from distantly related species. But to produce adaptation, something has to winnow out the wheat from the chaff: those variants that reduce reproduction from those that enhance it. And that's natural selection." Coyne seems to think natural selection is so powerful that the actual sources of variation aren't much worth mentioning.

Darwinians chafe over this low number of true believers in Evolution 1.0. They complain that science education is lacking; that a large percentage of Americans are superstitious and stupid; that fundamentalism and religious dogma have darkened the minds of millions of gullible citizens.

Religion Is Not the Only Reason People Reject Darwinism

Based on the many conversations I've had with people from all walks of life, I doubt any more than half the people who disbelieve evolution do so because they think their religion forbids it.

Let's talk about the rest of the crowd for a minute. Why are they unconvinced?

My own conversations with thousands of people have shown me that most Americans don't believe Evolution 1.0 because they don't buy the doctrine that it's accidental and purposeless.* Most people sense at a gut level that sophisticated systems and mechanisms don't endlessly upgrade without some sort of guidance. Something just isn't quite right.

We all have experience watching buildings being built, ideas evolving, and products being turned out into the marketplace. None of us has seen anything built solely from random events helped along by a process of elimination—ever. Yet that's the essence of traditional Darwinism. Zero design, zero purpose, and zero plan contradicts all normal everyday experience.

Atheist Richard Dawkins claims, "Evolution produces such a strong illusion of design it has fooled almost every human who ever lived" (115). But could the reason many people just don't buy the Richard Dawkins/Jerry Coyne version of evolution be, instead, that they smell an agenda? An agenda that has driven the atheist community to omit any scientific detail that reveals the purposeful, real-time adaptations of living things?

Is that why Dawkins and Coyne expound upon the wonders of natural selection, but don't bother to show you the mechanisms that generate new traits and originate new species?

---

* Alfred North Whitehead said, "Those who devote themselves to the purpose of proving that there is no purpose constitute an interesting subject for study."

And since it's politically correct to speak dismissively about God, and since it's not okay to raise uncomfortable questions in the public square, *theirs* is the version that ends up dominating entry-level science textbooks.

The Neo-Darwinists deny purpose, even though their language drips with purpose-laden terms like "selfish genes." They criticize Young Earth Creationists for insisting the Earth only *appears* to be millions of years old . . . yet they claim that living things only *appear* to be purposeful.

What's the difference?

When you allow nature to simply tell its own story, when you subtract randomness from the equation and replace it with the goal-seeking systems, evolution, in the form of Evolution 2.0, finally begins to make sense. You find that real-world biology doesn't support atheism at all. It speaks to a world that's even more amazing than most people dared to believe.

Don't Lie to Your Gut

All of us make decisions based on intuition and gut instinct that we cannot rationally defend in purely factual terms. Malcolm Gladwell's best-selling book *Blink* is precisely about that whole phenomenon, and in it he shows that this instinctive "blink" reaction is right more often than it's wrong.

I know gut instinct is no substitute for scientific rigor. But what Gladwell is saying is that gut instinct still sometimes turns out to be more accurate than mountains of allegedly scientific proof.

Remember the beginning of my story, where I asked, "Do the biologists know principles of design that they never bothered to teach me in engineering school?" The answer was . . . *well, sort of . . . but natural selection isn't one of them.*

It's more like cells have known things for millennia that humans suddenly discovered in the last 100 years. Cells have so many more things yet to teach us.

So far we've seen that information technology confirms what millions of people's gut instinct had been telling them since Darwin: *Garbage in,* *garbage out*. Philosopher Thomas Nagel (934) called Neo-Darwinism "a heroic triumph of ideological theory over common sense."*

Many times I've asked software coders what would happen if they tried to debug their programs by randomly corrupting millions of copies until one lucky one worked correctly. They chuckle and look at me like I'm from Mars.

My friends are probably a lot like yours. They've never studied evolution in detail. Many don't really care one way or the other.

I also have deeply religious friends who have no problem whatsoever if God used evolution to put us on the planet. Francis Collins, former director of the Human Genome Project and now head of the National Institutes of Health, formed the BioLogos foundation to advance his conviction that evolution is fully compatible with faith.

As I've shared my own discoveries with both friends and strangers, I've had an entirely different experience than Jerry Coyne with his business group, the guys who don't trust his 1.0 version of evolution.

As I explain how bacteria develop resistance to antibiotics, audiences listen with rapt attention. Just like Melanie, my friend Bob's daughter, did. Even her Young Earth Creationist dad thinks this is fascinating (even if he doesn't quite accept it on a grand scale). Nobody had ever told them any of this before.

Jerry's Business Breakfast Versus Perry's Business Breakfast

I spoke to a business breakfast in Chicago, too. My talk was to 130 business consultants on a Sunday morning in September. They had traveled there from all over the world—Australia, New Zealand, Spain, Greece, and Portugal.

In less than an hour I sketched the ABCs of Transposition, Horizontal Transfer, Epigenetics, Symbiogenesis, and Hybridization. I explained that the first three achieve stepwise, "kaizen† continuous improvement" adaptations, and the last two achieve "quantum leap" results.

---

* Thomas Nagel, an atheist, rejects Darwinism in his book *Mind and Cosmos: Why the Materialist Neo-Darwinian Conception of Nature Is Almost Certainly False* (934). He argues that materialism has no capacity to explain basic human attributes like belief and consciousness. He posits that there must be purposeful forces in the universe.

† *Kaizen* is a Japanese word that means constant, unrelenting improvement. Western business has adopted the idea so completely that kaizen is considered a standard English word.

I gave business examples, like a KFC/Taco Bell restaurant being a hybrid of two fast food chains. I drew parallels to business and told them, if they're looking to "dial in" degrees of innovation for their clients with a clear range of choices from slight improvement to radical reinvention, they need look no further than Mother Nature for powerful adaptation models.

When you encounter a situation where your existing business model has stopped working, I suggested, ask yourself: Should I rearrange existing departments? Should I copy some new components from a similar business? Could I simply switch off or switch on parts I already have? Might I merge two similar businesses? Or shall I bring some completely new element inside the existing one, like blue-green algae + protozoan = plant cell with photosynthesis?

In my talk I spent extra time on Symbiogenesis ("cooperative mergers"), because the way algae reinvents the cell to make a plant possible is beautiful. A Starbucks in the lobby of a Hilton hotel changes the entire *feel* of the first floor. It attracts visitors who aren't hotel guests and would never otherwise come to the property. "Intel Inside" advertises the computer's engine to the whole world. And the material on Symbiogenesis set the stage for another speaker who talked about merger-acquisitions. What's a leveraged buyout? It's when virus assumes control of host.

The Swiss Army Knife concept is *systematic creativity.* You don't have to wait for some random idea to suddenly spring on you from the ether. You can reach into your bag of 2.0 evolutionary tools and decide what innovation you want to apply, right now.

I explained how I'd been raised a Young Earth Creationist, and had initially been inoculated against these fascinating truths. I had to grow up and embrace a more nuanced understanding of God and nature.

I asked, "How many of you have ever heard *any* of this at all, anywhere else?"

Out of 130, maybe 20 hands went up. All was completely new to the rest. And these were not uneducated people.

As for the arms-crossed skepticism Jerry Coyne got? I didn't get that from *anyone.* Not even the Creationists in the crowd (I certainly had some). Many were furiously scribbling notes. One guy asked me to go over the five blades once again. A woman shook my hand after my talk and smiled, saying, "I'm Buddhist and my beliefs are much different than yours, but *thank you, thank you, thank you* for addressing this stupid war between science and religion. That was soooo good!" A devout Muslim couple said the same thing.

The Science of Systematic Creativity

There's an entire genre of literature on systematic creativity. Perhaps the best-known approach is called TRIZ (pronounced "trees") by Russian engineer Genrich Altshuller. TRIZ is a 40-step Swiss Army Knife for solving engineering problems, especially for physical products and devices. The book *40 Principles: TRIZ Keys to Technical Innovation* reports: "Scientists claimed that inventions were the result of accidents, mood or 'blood type.' Altshuller could not accept this—if a methodology for inventing did not exist, one should be developed... Invention is nothing more than the removal of a technical contradiction with the help of certain principles" (701).

Altshuller's efforts to empower Russian engineers with his method for systematic creativity included a letter to Josef Stalin. The end result of his contributions was that he and his partner, Rafael Shapiro, were charged with "inventor's sabotage" and sentenced to 25 years' imprisonment (701). Have you ever noticed that the people who bark loudest about "science" and "progress" are often the bitterest enemies of both?*

---

* As mentioned in chapter 1, the Salem Hypothesis states: "An education in the Engineering disciplines forms a predisposition to Creation/ID viewpoints" (256). Medical doctors and dentists have this same bias. Indeed, the first backer of the Evolution 2.0 Prize was a nationally renowned surgeon. By now you know exactly why engineers, doctors, and programmers are skeptical of Darwinism: Nobody uses random mutation and natural selection to design a product, and no healthy organ in the human body looks like an accumulation of random accidents. The Neo-Darwinian theory violates fundamental engineering principles like information entropy, then replaces empirical proof with futile speculations that you could never test even if you wanted to. The hard-working professionals who brought us the Information Age know this only too well. As an electrical engineer who's accustomed to theoretical models being accurate to better than 1 percent, I'm not merely skeptical of Darwinism; I'm appalled by it. How unjust it has been to sweep astonishing insights like Symbiogenesis under the rug in order to hold the random mutation hypothesis in place. Coming full circle, though, an accurate understanding of evolution gives engineers many new useful tools. Nothing would benefit medicine and technology more than a rigorous understanding of the cell and how it adapts to changing threats.

Why Should Evolution and Design Be an Either/Or Proposition?

ONLINE SUPPLEMENT

Systematic creativity tools for "Innovation On Demand"

www.cosmicfingerprints.com/supplement

Why can't we have both? Why shackle science with a two-party system? Evolution doesn't have to be competitive; it can be cooperative. If evolution is purposeful... if all codes need designers... then the universe is a purposeful place, not a blind, pitiless, indifferent place. Evolution 2.0 saves you from the bitter, nihilistic corset of Darwinism. Science and faith need not be at war.

When you declare, "There's no reason for religion and science to be at war with each other," the majority of people are receptive. The debate is not an esoteric ivory-tower conversation. It's real, and road-tested Evolution 2.0 causes regular folks to sit up and listen every single time. Even at casual dinner parties.

Oh yeah, and many people put their fork down when they hear how many of the greatest discoveries made in science—responsible for all the natural laws of our universe—were made by men of faith. What, really?

CHAPTER 28

On the Shoulders of Giants: When Men of Science Were Also Men of Faith

We are of the spirit,
Truly of the spirit
Only can the spirit
Turn the world around

—HARRY BELAFONTE

EVER WONDER WHERE we got science in the first place? It's because many very smart people hundreds of years ago believed that God made an orderly, structured universe—and science was the way to show this. In short, science was conceived as the study of the mind of God (927).

Did you know that most of the greatest scientists of the 1500s, 1600s, and 1700s were devoutly religious? Even as early as the year 1000, Pope Sylvester II was the leading mathematician and astronomer of his day. He rose from humble beginnings to the highest office in the Christian Church "on account of his scientific knowledge" (903). His innovative abacus has been dubbed the first computer, and one scholar called him "the Bill Gates of the end of the first millennium" (903).

I have a well-educated friend who's a history buff, who was surprised by this. She told me we've been sold such a bill of goods these days that science and faith are not related, and hearing that the Founding Fathers of Science were men of faith was a revelation to her.

It's time people heard the truth: Modern science was birthed from the belief that to discover and quantify the order of the natural world was an act of worship.

Listen to this from the mouths of the great science pioneers themselves...

1543 COPERNICUS, ASTRONOMER: "[I]t is [the philosopher's] endeavor to seek the truth in all things, to the extent permitted to human reason by God" (907).

1596 JOHANNES KEPLER, ASTRONOMER: "Geometry is unique and eternal, a reflection from the mind of God... The diversity of the phenomena of nature is so great, and the treasures hidden in the heavens so rich, precisely in order that the human mind shall never be lacking in fresh nourishment" (931).

1615 GALILEO GALILEI, ASTRONOMER: "I do not feel obliged to believe that the same God who has endowed us with sense, reason, and intellect has intended us to forgo their use" (917).

1674 ROBERT BOYLE, THE WORLD'S FIRST MODERN CHEMIST: "All [is] upheld by His perpetual concourse, and general providence; the same philosophy teaches that the phenomena of the world, are physically produced by the mechanical properties of the parts of matter; and, that they operate upon one another according to mechanical laws" (902).

1700 ISAAC NEWTON: "It is the perfection of God's works that they are all done with the greatest simplicity. He is the God of order and not of confusion" (936).

1931 ALBERT EINSTEIN: "The most beautiful and most profound experience is the sensation of the mystical. It is the sower of all true science. He to whom this emotion is a stranger, who can no longer wonder and stand rapt in awe, is as good as dead" (913).

1931 MAX PLANCK, WHO WON THE NOBEL PRIZE FOR QUANTUM THEORY: "Both Religion and science require a belief in God. For believers, God is in the beginning, and for physicists He is at the end of all considerations... To the former He is the foundation, to the latter, the crown of the edifice of every generalized world view"(939).*

1960 WERNER HEISENBERG, NOBEL PRIZE–WINNING FATHER OF QUANTUM MECHANICS: "The first gulp from the glass of natural sciences will turn you into an atheist, but at the bottom of the glass God is waiting for you" (951).

2007 FRANCIS COLLINS, U.S. DIRECTOR OF THE HUMAN GENOME PROJECT: "Will we turn our backs on science because it is perceived as a threat to God, abandoning all the promise of advancing our understanding of nature and applying that to the alleviation of suffering and the betterment of humankind? Alternatively, will we turn our backs on faith, concluding that science has rendered the spiritual life no longer necessary, and that traditional religious symbols can now be replaced by engravings of the double helix on our altars?

"Both of these choices are profoundly dangerous. Both deny truth. Both will diminish the nobility of humankind. Both will be devastating to our future. And both are unnecessary. The God of the Bible is also the God of the genome. He can be worshipped in the cathedral or in the laboratory. His creation is majestic, awesome, intricate and beautiful—and it cannot be at war with itself. Only we imperfect humans can start such battles. And only we can end them." (906)

---

* Planck also said, "Under these conditions it is no wonder, that the movement of atheists, which declares religion to be just a deliberate illusion, invented by power-seeking priests, and which has for the pious belief in a higher Power nothing but words of mockery, eagerly makes use of progressive scientific knowledge and in a presumed unity with it, expands in an ever faster pace its disintegrating action on all nations of the earth and on all social levels. I do not need to explain in any more detail that after its victory not only all the most precious treasures of our culture would vanish, but—which is even worse—also any prospects at a better future" (939).

Peter Hitchens, brother of the late Christopher Hitchens, echoes Planck's observation many decades later in his recent book *Rage Against God* (925). He notes that in the Soviet Union, state-enforced atheism had stripped Russian culture of all its gentleness, trust, and dignity. He lived in the USSR for two years, and when he would hold doors open for commuters in the subway, they would scowl at him, believing he must be playing some kind of trick. How is the modern prohibition against mentioning God in science discussions anything other than a milder form of state-enforced atheism?

History illustrates that a belief in God does not impede the advancement of scientific discovery, since the titans of science like Newton, Galileo, Bacon, Kepler, Descartes, Boyle, Faraday, Mendel, Kelvin, and Planck believed in God. Their faith was a fire that ignited their research.

Wisdom of Solomon 11:21, which is in the Catholic Bible (the Apocrypha), says in part, "Thou hast ordered all things in weight and number and measure." The core assumptions of the scientific worldview trace back to Jewish theology. The book was traditionally attributed to Solomon, who lived 3,000 years ago; even though Solomon was probably not the author, it was written no later than 100–200 B.C. (920).

I am aware of no older book that makes such a definitive statement of the rationality and measurability of the universe; the seeds of earliest science originate in the Bible. I also tell my Protestant friends that if they haven't read the Apocrypha, they're missing out on fascinating stuff.

Science and Religion Are Actually Inseparable

Albert Einstein's belief that it was unlikely that the universe was created without some higher power motivated his interest in science. He once remarked, "I want to know how God created this world. I am not interested in this or that phenomenon, in the spectrum of this or that element. I want to know His thoughts, the rest are details" (928).

Even Darwin makes extensive theological arguments in *On the Origin of Species* to support his theory: assumptions about what kind of world a theistic or deistic God would or would not make. So much so, that if one were to separate religion from science, Darwin's book could not reasonably be permitted to be studied in science classrooms.

Stephen Dilley explores this in his paper "Charles Darwin's Use of Theology in the *Origin of Species*" in the *British Journal for the History of Science*. Several theological assertions are central to Darwin's case. Dilley points out that Darwin believed life was originally created by a deistic god who created the universe, then abandoned it to the outworking of fixed natural laws ever after (912).

In large part, science rests on an unprovable assumption (some might even propose the word *faith*) that the universe is governed by fixed, discoverable laws; that it operates without the need for constant tinkering by the creator; that the universe has a degree of freedom to follow its own course.*

While it is possible and often desirable to debate science without explicitly bringing religion into the discussion, it is impossible to do so without invoking the "big questions." The fact is, *everyone* brings theological assumptions—even if those assumptions are about the non-existence of God—to the table. It's just a question of whether they're forthright about it or not.

People who claim that science and religion are "non-overlapping magisteria" as Stephen Jay Gould said (618) are trying to have their cake and eat it, too. That's because Gould's assertion that science and religion do not overlap is not a scientifically testable statement; it's a religious and philosophical assertion.

So why the raging battle between faith and science?

Actually...I Hold Christians Responsible for This

I'm a Christian, so I can say that. If you're a Christian and you think science and faith are enemies...I love you, brother or sister, but in good conscience I have no choice but to loudly object. G. K. Chesterton said, "The Christian is quite free to believe that there is a considerable amount of settled order and inevitable development in the universe. But the materialist is not allowed to admit into his spotless machine the slightest speck of spiritualism or miracle."

Why do people today believe faith and science are enemies? Especially when science itself relies on so many unprovable metaphysical assumptions—like the regularity of nature.

Some people blame atheists and people who hate religion for the current Great Divide between Faith and Science. I admit, their desire to stamp out religion is considerable.

But too many people of faith have abdicated their responsibility (once their noble legacy) to embrace scientific inquiry. They have thrown the door

* The Big Bang theory was first put forth by Belgian Catholic priest and physicist Georges Lemaître in 1927. He said, "Scientific progress is the discovery of a more and more comprehensive simplicity...The previous successes give us confidence in the future of science: we become more and more conscious of the fact that the universe is cognizable" (921). Lemaître was harshly criticized for denying that the universe was infinitely old, and his detractors charged that he was trying to support Aquinas' arguments for God. Atheist Fred Hoyle hated the idea so much he called it the "Big Bang" and the name stuck.

wide open for skeptics to preach a purposeless universe.* To non-Christian people, I apologize for Christians who at times have misrepresented or ignored science because they distrust it.

The Made-Up War Between Science and Religion That Became Real

The alleged war between faith and science has a name. It's called the *Conflict Thesis*. It's a substantial collection of stories that depict everyone who lived between 300 A.D. and 1500 A.D. as ignorant, superstitious, fearful, and opposed to progress. A prime example is the famous story of how people in the Middle Ages thought the Earth was flat, so the queen of Spain warned Christopher Columbus that he might sail off the edge of the Earth. It isn't true.

No educated person in the Middle Ages ever thought the world was flat. The story was made up by chemist and photographer John Draper in 1874, who fabricated it to make Catholics look bad (915). It was part of a larger modern myth that people in the Middle Ages were finally liberated by science's triumph over darkness, dogma, and superstition.

Modern historians consider the term *Dark Ages* to be inappropriate and misleading because the so-called Dark Ages never happened (947). Toilet paper, distillation, high-purity glass, the hang glider, chemotherapy, metal block printing, oral anesthesia, the pinhole camera, the programmable analog computer, the torpedo, and eyeglasses were all invented between 500 and 1300 A.D., along with hundreds of other inventions (919).

The ancient Greeks didn't invent science (944). Science, technology, and standards of living steadily improved from the fall of the Roman Empire to the Renaissance (943) and the Church never tried to outlaw operating on human cadavers (948), though Greek and Roman religions did (see 200).

---

* People from *every* corner of this debate have taught me important and valuable things. Werner Gitt is a Young Earth Creationist, but that doesn't keep his book *In the Beginning Was Information* (310) from being the goldmine of insights that it is; it makes his approach more iconoclastic. Atheist Jerry Coyne's book *Why Evolution Is True* (105) omits all the most interesting facts about evolution, and overall is quite misleading. I can't exactly say I enjoyed the read. Nevertheless it still presents useful evidence *for* evolution. The most important thing I have learned is that both sides have much to teach each other. Faith and science can stop being enemies now.

"In the Dark Ages, contrary to what most people think, science was central to the lives of monks, kings, emperors, and even popes. It was the mark of true nobility and the highest form of worship of God."

—From "Everything You Think You Know about the Dark Ages is Wrong" by Nancy Marie Brown in an interview based on her book *The Abacus and the Cross: The Story of the Pope Who Brought the Light of Science to the Dark Ages* (903).

People who said, "Nature could never have possibly done that. Only God could do that" rolled out the red carpet for those who insist, "With enough stars and billions of years, anything can happen . . . all you need is a happy chemical accident."

As soon as you declare that evolution is random, you've hit the end of the scientific road. So we have two sides that have reached dead ends, who in their frustration blame each other.

Belief in God Does More Than Drive Science Forward; It Also Guards Against a Scientific Dystopia

Science and technology grant a tiny minority of human beings great power over the many. If there were no God, if evolution through natural selection were an inevitable feature of an utterly indifferent universe, we would have no rational basis whatsoever for racial equality, morality, and human rights. All such notions would be nothing more than our frail emotional reactions to the hard realities of natural selection.

This is why Darwinism is inextricably associated with ethnic cleansing, racism, eugenics, and genocide. Darwin's original full book title was, after all, *On the Origin of Species by Means of Natural Selection, or the Preservation of Favoured Races in the Struggle for Life*. In it, Darwin wrote, "The civilised races of man will almost certainly exterminate, and replace, the savage races throughout the world" (108). Remarking on social institutions that care for the poor in his later book, *The Descent of Man, and Selection in Relation to Sex*, Darwin wrote, "Excepting in the case of man himself, hardly any one is so ignorant as to allow his worst animals to breed" (109).

Racism, eugenics, and genocide are *totally logical* if Darwin's understanding of humans is correct and we are not spiritual beings.

Darwinism is racist and inhumane. Richard Dawkins acknowledges this in his assessment of theistic evolution views of man:

> In plain language, there came a moment in the evolution of hominids when God intervened and injected a human soul into a previously animal lineage (When? A million years ago? Two million years ago? Between *Homo erectus* and *Homo sapiens*? Between "archaic" *Homo sapiens* and *H. sapiens sapiens*?). The sudden injection is necessary, of course, otherwise there would be no distinction upon which to base Catholic morality, which is speciesist to the core. You can kill adult animals for meat, but abortion and euthanasia are murder because human life is involved. (900)

Man is special precisely because "God breathed the breath of life into the man and he became a living being" (Genesis 2:7). This is why it's okay to eat a cow and it's not okay to eat your office manager. This is why we humans ask ourselves questions that animals don't appear to trouble themselves with. It's why humans are irrepressibly religious.

We are, at our core, spiritual creatures. Humans inherited a craving to transcend the rules of Darwinism. That's why so many people bristle at the Darwinian worldview. Most of us cringe at how even artists and musicians have to claw their way to the top of some heap just to find fans to enjoy their work. Have you ever noticed that at funerals, the eulogy is always about how loving, giving, and "non-Darwinian" the dearly departed was?*

---

* Frank Schaeffer, the author, director, and screenwriter and son of the famous theologian Francis Schaeffer, said in an interview, "Most people don't really want to live only according to narrowly defined material facts. Most of us try to direct our human primate evolutionary process along ethical non-material lines. We impose standards that do not come from nature. Nature is cruel yet we try not to be. We prosecute people for war crimes that are no more destructive than what happens every day in the churning cauldron of life where everything is eaten and where death is the only incubator of life. We call murder wrong although it's the most natural thing on earth.

"We've decided to let an imagined utopian ideal, a future Eden if you will, rule our present despite this being a spiritual non-material-universe-based choice that flies in the face of natural selection. We impose ethics that exist only in our heads upon the material universe. We are part of nature yet we have decided to be nicer than nature. There would be no war crimes trials unless our ethically evolved selves questioned the method of evolution itself." (940)

Without God, "human rights" is an idea that hangs in midair with no higher law, no external support, forever vulnerable to some arbitrary cost-benefit analysis. Belief in God is essential grounding for guaranteeing human rights.

In a godless universe, there is no moral authority. There is only *my* feelings versus *your* feelings, *our* feelings versus *their* feelings. Those feelings are nothing more than chemicals coursing through our veins. Eventually, differences get settled with bloodshed. "Might is right." A spiritual foundation keeps science in check. The track record of the greatest atheist societies in history—Mao's China, Lenin and Stalin's Russia—is well known. The 20th century offers warning enough: Throw God under the bus at your extreme peril.

God Is No Longer Banned from Science. So Let's Take Nonrandomness a Step Further: A Fully Scientific Hypothesis

When you replace the Darwinian random mutation theory with that of goal-seeking mutations, you've taken the most important step toward correcting the dogmas of Darwinism.

Mutations aren't random, they're goal directed (643, 645, 664).

Natural selection has zero creative power; it's only the final step of elimination after organisms have performed magnificent feats of genetic engineering.

Evolution, as detailed in Evolution 2.0, isn't gradual; the majority of measurable progress occurs in short periods of time, followed by long periods of general stability.

Evolution 2.0 events reverse information entropy. The actions cells take to communicate with each other, edit their genomes, engage in symbiotic relationships, exchange DNA with other cells, and form hybrids increase information and order in the universe. Novelty comes from the cells themselves.

All of which is to say, Evolution 1.0 is backward, broken, and blatantly contradicts the most important known facts about biology.

A thoroughly scientific hypothesis presumes order and structure wherever it is reasonably warranted. Therefore:

- When a letter or codon in DNA changes and confers a positive benefit to the organism, we assume the mutation occurred in response to inputs from the environment, not randomly.

- When code moves from one place to another, we do not assume it just accidentally jumped to that location for no reason at all. We assume there is a process that explains why the cell put it there.
- We know cells are cognitive and are able to communicate with other cells, the same way animals communicate with each other. Thus we assume that when editing their DNA, cells, like humans, make calculated estimates, not random guesses.
- You didn't get to work this morning by flipping a coin every time you hit an intersection. Humans only flip coins either when we explicitly want to make a random choice, or else because we simply do not know or don't care. We assume cells operate the same way. Even in the immune system, cells generate targeted permutations from a library of combinations. They don't just randomly slap molecules together.
- If a cell receives genetic material from another cell through Horizontal Transfer, we assume it did so with some bias toward a desirable outcome, not haphazardly.
- We assume cells behave somewhat like humans and animals: They make the best guess they can based on the limited information available. Just because they operate systematically doesn't mean they always make the right decision, or even that they usually do. Like us, they fail more often than they succeed. But in aggregate, they make progress and they share their progress with their community.

Let's try a new rhythm: Let's propose that God (or a supremely powerful being, if that makes you more comfortable) made a rabbit hole so deep, we don't know how far it goes. We only know there's always more to discover.

And that's how we break the deadlock between Darwin and Design.

BULLET POINT SUMMARY:

- Neo-Darwinism says Random Mutation + Natural Selection + Time = Evolution.
- Random Mutation is noise. Noise destroys.
- Cells rearrange DNA according to precise rules (Transposition).
- Cells exchange DNA with other cells (Horizontal Gene Transfer).
- Cells communicate with each other and edit their own genomes with incredibly sophisticated language.
- Cells switch code on and off for themselves and their progeny (Epigenetics).
- Cells merge and cooperate (Symbiogenesis).
- Species 1 + Species 2 = New Species (Hybridization). We know organisms rapidly adapt because scientists produce new species in the lab every day.
- #Evolution in 140 characters or less: Genes switch on, switch off, rearrange, and exchange. Hybrids double; viruses hijack; cells merge; winners emerge.
- Adaptive Mutation + Natural Selection + Time = Evolution 2.0
- DNA is code. All codes whose origin we know are designed.
- Where do codes and linguistic rules of DNA come from? Evolution 2.0 prize.
- Answering this question will produce billion-dollar medical and technological breakthroughs.
- Darwinists underestimate nature. Creationists underestimate God.
- Man yearns to escape Darwinism and embrace equality and human rights. Those are spiritual values, not scientific principles. That's why it's time to end the war between science and religion.

CHAPTER 29

Why So Much Pain and Suffering in the World?

Pain and suffering
I am destruction
The pain, the pain . . .
Human race, you're going to writhe now

—Iggy Pop

I'VE GOT A FRIEND NAMED JESS. She used to live a couple miles from my place. Life has been HARD on Jess.

Several years ago her husband Jamie contracted a lethal form of leukemia. After a two-year battle she lost him.

Then her 11-year-old son, Alex, started developing bruises on his skin. A doctor's appointment revealed that he, too, had leukemia.

Wow. An 11-year-old boy with leukemia. Imagine facing that.

Alex went through nine months of brutal chemotherapy treatments. The disease went into remission.

Eighteen months later it was back with a vengeance.

More chemotherapy. More prayers. More desperation.

Alex chose to stand up and FIGHT.

Since Alex was losing his hair, several of his friends, including his best friend Dylan and my son Cuyler, decided to shave their heads as a sign of solidarity with Alex during his healing and recovery process.

Alex began his chemo treatments. Dylan's shaved head lightened up Alex's first day of chemo.

Two years after the saga started, an infection raged out of control and Alex died, 13 years young.

Jess had now lost both husband and firstborn son to leukemia.

There wasn't much happy about that memorial service. She made it through Alex's funeral under the care of friends, desperate prayers, and a couple pints of tequila.

We had all yearned for Alex to be healed. For this curse to be lifted. But Alex lost. Deliverance did not come.

During those brutal Chicago winter days, when Jess grieves for what she's lost, God seems a trillion miles away. *You call that a loving god? Any deity who'd create a world like this sounds like a blundering tinkerer who winds up watches and leaves them cracked and rusting on the cold wet ground.*

Philosopher David Hume, in his *Dialogues Concerning Natural Religion*, offered, I think, his *real* reason for not believing in God:

> Were a stranger to drop on a sudden into this world, I would show him, as a specimen of its ills, a hospital full of diseases, a prison crowded with malefactors and debtors, a field of battle strewed with carcasses, a fleet foundering in the ocean, a nation languishing under tyranny, famine, or pestilence. To turn the gay side of life to him, and give him a notion of its pleasures; whither should I conduct him? To a ball, to an opera, to court? He might justly think that I was only showing him a diversity of distress and sorrow. (926)

I am pretty sure that most people's reason for rejecting purpose in nature has not so much to do with science, and much more to do with the frustration, disappointment, and rage at a world of such intense pain and suffering.

The Darwinian doctrine that nature is purposeless and random *appears* to relieve us from answering questions like, "Why do 13-year-olds die of leukemia?" or "Why is there death and disease?" or "Why are there birth defects like Down syndrome and spina bifida?" and so many more.

Could it be that skeptics often reject God because they can't resolve the tension of pain and suffering with an omnipotent being? Some

Creationists reject evolution for nearly the same reason—because they can't imagine God making a world that includes death from the word "go."

The pristine utopia of a Young Earth is so much simpler and more palatable than the untamed layers of the paleontologist. But I came to realize that the Genesis command to "replenish the earth and subdue it" meant that even in Eden, there would be much work yet to do (see appendix 2).

Code's inference to a designer transports us straight to the outer edge of science. It demands that we ask the Big Questions about God, philosophy, and the metaphysical world. I said at the beginning that this is not a religious book and I'm not going to attempt to answer these questions here. (You can explore this on my website, if you wish, at www.cosmicfingerprints.com.)

Science itself does not answer these questions, but it does validate them. It shows us that codes are always the result of purpose, and that purpose has to come from somewhere. Indeed, science ushers us right to the entrance of the philosophy and theology departments. It asks questions that beg to be answered.

How can a sane person *not* ask, "What sort of perfect God creates an imperfect world? What kind of logic is that?"

There is a widespread belief that discussions of religion and philosophy should be banned from the science classroom and the laboratory. That way of thinking doesn't engender a spirit of inquiry.

If we can discuss science in history and philosophy classes, then . . . why aren't we allowed to discuss history and philosophy in science class?

I lost my dad during high school. He died of cancer when I was 17. He was 44. I faced that same question: Does God take the good guys and abandon the rest of us to fend off the bad guys all by ourselves?

I had two choices: I could chalk up dad and my memories of him as the result of so many billiard balls banging around in the universe, blindly producing some result or another. I could ascribe whatever feelings I had about that to nothing more than stupid chemical reactions in my brain.

Or, like Jess—whose faith and whose connection with the Divine grew through her loss—I could swallow the hard pill. In the battle between good and evil, evil had prevailed.

You can choose to deny there's any purpose at all. But there's no success or failure without purpose.

You can't fight for good and against evil until you acknowledge the existence of both. A purpose allows us to measure the way things are against the way things should be.

The good is real. The bad is real. I chose to accept the good and the bad.

I don't know why the world is the way it is. What I do know is that despite fires and floods, ice ages and meteors and famines, life is still here and it still thrives.

Olympic skaters dazzle the world with their choreography, composers write symphonies, and relief workers place AIDS orphans in loving foster homes.

Tiny cells rewrite their code with an ingenuity that puts the world's smartest programmers to shame.

I had to outgrow my youthful notions of God. The six-day Creationist God I grew up with gave way to a richer conception of the Divine, that of a master programmer who could spin a strand of code that fills the whole Earth with beauty. A God who seems to be much more interested in processes and wisdom, maturity and growth, than quick fixes.

The God I imagined as a child yielded to one who weaves a complex story, a story not just for children but for adults, a grand epic tale. A God who invested himself into that tale by creating human beings in his image. Who endows us with a spirit that feels the joys and the heartaches of that tale. One who beckons us to live in the tension of paradox and mystery.

You can tell yourself stories of junk DNA and vestigial organs, or you can ask why those things are there. You can criticize from the sidelines, or you can get in the game and do something great for somebody else.

But you can't do both.

None of us will ever get through life without facing the Big Questions. And it is no longer possible to use the remains of 20th-century Darwinism to dodge them, either. If you struggle with the Big Questions, now is the time to lace up your boots and commence your journey of answering them. The science knowledge you've gained in this book won't salve your wounds, but it will assure you that your search for meaning rests on solid ground.

CONCLUSION

CHAPTER 30

Brother Bryan Comes Around

I finally see the dawn arriving
I see beyond the road I'm driving
Far away and left behind
Don't look back

—BOSTON

MY BROTHER BRYAN had been watching my online science debates with keen interest. He and I debated the questions in this book at length—for years.

Early in our discussions, Bryan would say to me, "Perry, how can you blame scientists for not liking your conclusion that DNA is designed? It's their responsibility to look for and find a naturalistic cause. If they simply attribute it to God, they're abdicating their job as scientists."

I would respond, "Yeah, I totally respect that, and you are correct. It's their job to find a naturalistic, purely scientific explanation for the origin of life. But what if they can't? If scientists deny the evidence that points to the genetic code being the product of an intelligent coder, that's abdicating their job, too!"

And although he was slow to embrace my conclusion, Bryan couldn't fail to notice the blowback I got from atheists even when they couldn't counter my evidence.

Driving around in the car one day, Bryan finally asked me, "How come they couldn't just admit that they didn't know?" He paused, looking out the window. "You know, all those atheists you're arguing with are just devotees of a different form of fundamentalism.

"Thank you, bro, for not allowing me to become an atheist."

At the beginning of this book, I told you about the promise I made to myself: *I am going to get to the bottom of this. Even if it costs me* everything. *I just want to know what's true.*

And now, after years of research, expense, scrutiny, and debates, my conclusion is: *Not only is Evolution 2.0 the most powerful argument* for *a Designer that I've ever seen (!), but people of faith were on the cutting edge of science for 900 out of the last 1,000 years. The rift between faith and science might heal if everyone could see how evolution actually works.*

When I began this journey, I was teetering. I was playing the part of someone who seemed to believe in God, saying all the right things to my friends, even as I was secretly doubting the whole entire thing. But now I had witnessed such awe-inspiring levels of order—cellular engineering feats light years beyond human imagination—and I knew: *only Someone or Something very great could pull this off.*

As for Bryan, he's still sorting out his faith questions. He and I both agreed that demanding a materialistic explanation for the Origin of Life was sort of like driving west toward the sunset expecting that you would eventually reach the sun. It might appear as though it's only 100 miles away, but think again—it's really 93 *million* miles away and you'll need a spaceship, not a car.

Similarly, information and consciousness require a different vehicle of explanation. Matter and energy don't tell you where they come from. They have to come from somewhere else.

Meanwhile, evolution itself—the *real*, 2.0 version, as opposed to the just-so story you read about in bookstores—continues to fascinate Bryan and me. Each of us has found ways to apply the concepts in our business endeavors. We've even started to use geeky words like *Symbiogenesis* with each other when we talk about technology mergers and cooperating companies.

Bryan's the president of my firm, and in the business we run together, we began borrowing ideas from Evolution 2.0 and applying them to business and marketing problems. Sometimes the only language we could find used terms borrowed from biology. (We especially like the term *Hybridization*—so very useful in business!) We even developed a tool for business owners called the Swiss Army Knife, a process for identifying emotional hooks and generating systematic variations in online ads.

Not only is Evolution 2.0 fascinating, it turns out to be practical—because it's applicable to almost any business anywhere! Each of us would describe adaptive mechanisms to our friends and colleagues. Without exception, those people found it fascinating. Even 12-year-olds like Melanie at the beginning of this book.

CHAPTER 31

"Why Should I Care and What Should I Do?"

What you gonna do with all that junk?
All that junk inside your trunk?

—BLACK EYED PEAS

SUPPOSE YOU HAD a foreign car—you know, one of those funny ones where the engine is in the trunk. You took it to the mechanic, and your mechanic phoned you to say, "I found some junk—like, a big cluster of hoses in your engine—that appear to be useless. Do you want me to take them out?"

You'd probably panic. Then you would (hopefully) say, "I'm pretty sure that if Peugeot put those hoses in, they must be there for a reason. So no, sir, you cannot take those hoses out. I will pick up my car tonight and find another mechanic."

If you don't have knowledge of (or respect for) something, aren't you more inclined to dismiss it as worthless?

I could rant about the times that Darwinism has been employed to justify eugenics and exterminate millions of people by communist regimes (932). Much has been written about the persistent connection between Darwinism and racism. The Nazis used genetics as a justification; the Communists had Lysenko (909). But my concern (in this book, anyway) is for science, not the twisted application of science and the grotesque machinations of deranged world leaders.

How much damage has been done to *science and medicine* by Darwinism?

Take the issue of so-called junk DNA, or "noncoding DNA" as it's known today. We are aware that 3 percent of the human genome codes for proteins. In 1972, a scientist (315) coined the term *junk DNA* to describe the 97 percent of DNA with no known function. Some scientists still maintain that large portions of the genome are useless accretions of evolutionary garbage.

ONLINE SUPPLEMENT

RIP Junk DNA, 1972–2012

www.cosmicfingerprints.com/supplement

The ENCODE project ("Encyclopedia of DNA Elements") was started in 2003 to find all the functional elements of the human genome. The *New York Times* announced, "Bits of Mystery DNA, Far from 'Junk,' Play Crucial Role," and went on to say:

> The human genome is packed with at least four million gene switches that reside in bits of DNA that once were dismissed as "junk" but that turn out to play critical roles in controlling how cells, organs and other tissues behave. The discovery, considered a major medical and scientific breakthrough, has enormous implications for human health because many complex diseases appear to be caused by tiny changes in hundreds of gene switches. (629)

Science magazine's report was entitled, "ENCODE Project Writes Eulogy for Junk DNA" (649). There is no such thing as junk in the trunk when it comes to DNA.

A tiny contingent of diehard junk-DNA advocates, such as Larry Moran of the University of Toronto, insisted that the ENCODE announcement was a "media fiasco" (230). But ENCODE's findings were unambiguous: At least 80 percent of our DNA is active and necessary. If you deleted it, our bodies would fail. Or our children or grandchildren would be missing something critical that they need to survive.

Do Larry Moran and other junk-DNA advocates also happen to share any particular bias with respect to religion? Check and see for yourself.

If we assume purposelessness in evolution, as is done in the 1.0 version, it's logical to expect a lot of junk. If we assume a designer, we assume there's a purpose to its inclusion, and therefore look into it until we discover it's not really junk after all.

In 2009, virus expert and physician Frank Ryan asked veterinary cell biologist Rachael Tarlinton why research into horse and cattle retroviruses was so scant. She replied:

> I've just submitted a grant application to look at the equine genome just to look at and characterize the actual retroviral load. You can access data for humans and mice, which have been very well studied, so we know what retroviruses are there, what their point mutations are, whether they are potentially able to produce proteins or not, but for other species that just does not exist... if you go and look at the genome map on GenBank, retroviruses and repetitive elements aren't annotated.
>
> If you try to search the genomes, they are actually excluded from the genome searches, because they are considered non-functional. Not interesting. Not important. Everyone is focusing on the [vertebrate genes that translate to] proteins and that's only 1.5 percent of the whole picture. (652)

If it bothers you that this research is being opposed, thank the junk-DNA crowd.

The people who say parts of DNA are junk say so out of ignorance, not knowledge. They don't know how to build a cell or a genome. They don't know what everything does. The burden of proof that junk DNA is truly junk is on them. Until they understand everything and can explain every nuance of the genome's operation in precise detail—until they can build a cell from scratch—their job is not done. The dictionary tells you why:

> **sci·ence**: The intellectual and practical activity encompassing the systematic study of the structure and behavior of the physical and natural world through observation and experiment.

Any scientist who takes his work seriously has no choice but to say, "I don't know what its function is, but my job is to fully engage in the systematic study of the structure and behavior of this until I do. So until I have a complete working model that describes the entire system in exact detail, I have no right to assume these stretches of DNA are junk."

The next time someone tries to tell you most of our DNA is junk, ask them this: "May I have permission to delete 50 percent of *your* genome? I promise to only delete the parts that you consider to be junk."

If you believe the universe is blind chaos, that's what you'll see when you look in the microscope. People see what they expect to see. The term *junk DNA* reflects the state of mind of some researchers, not the reality of what's in our cells.

This is not the first time Darwinism has vandalized science. There's been a long history of closing entire wings of genetic libraries and disparaging valid research programs using debased language, from Barbara McClintock's colleagues laughing at Transposition to Richard Dawkins dismissing Epigenetics to Jerry Coyne proclaiming that natural selection is "the only game in town." The time has come to stop this theft of valuable knowledge; it's a crime against science and humanity.

Another case in point is the matter of vestigial organs, the organs-in-your-body version of junk DNA. There is no such thing as useless organs; every organ in our body has a function, even if, like those whale legs, Evolution 2.0's Swiss Army Knife is saving it for a rainy day. Yes, even that troublesome appendix. (It's a "safe house" for symbiotic bacteria. The immune system uses it to host allies in its war against disease [605].)

But there's good news. Frank Ryan reports in his talks on the amazing role that viruses play in assisting adaptation:

> I lecture very widely these days, to doctors, geneticists, evolutionary biologists and to molecular biologists, and in every case where I am speaking to a "virgin" audience the reaction is the same—something bordering on astonishment. In practice, it is not difficult to promote understanding since the evidence base is now overwhelming. So much so that a single lecture is often all that it takes to educate colleagues from any of the biological or medical disciplines. The lectures do not end with disagreement or condemnation, but rather with a very high degree of interest and requests for more formal information, such as scientific papers. (652)

I've had the same experience with my own friends and colleagues. Even folks who were initially wary of Evolution 1.0 find Evolution 2.0 fascinating. Even if they don't agree with some of my views, they realize nature is even more amazing than they thought.

Suppressing Cancer Research

Just as Darwinian thinking has labeled so much of our DNA as junk, it defines cancer cell activity as "random." This postpones any cure.

You can't predict random behavior, so if cancer's next move is random, the problem is unsolvable. Not exactly helpful in the pursuit of cancer treatments. But the active mutations of cancer cells are *not* random. They're calculated responses to your body's immune system. The DNA changes are different for different types of cancer. I believe this is because the cell's Swiss Army Knife responds *in context* as it seeks to proliferate.

The Darwinian approach of "randomness" has blinded countless researchers to new models that might predict what cancer cells do. We all pay a price for that.

Science always presumes underlying order and structure. Our bodies are the triumph of millions of years of stunning genetic innovations, mergers, and partnerships. The world is what it is because of ingenious systems and designs. Not randomness. Not luck.

Humans destroy the Earth; cells rebuild it. Cells are smarter than humans. Cooperation trumps survival of the fittest.

I Want Everyone to Know About This

My conviction is that proper understanding of Evolution 2.0 will open the door to innumerable breakthroughs, in medicine, technology, and beyond. I'm inviting you to join the growing assembly who embrace Evolution 2.0 because they are deeply dissatisfied with the Darwinian status quo.

We're no longer okay with pretending the last 50 years of research never happened. We're tired of the scientific vandalism. We want to know how cells engineer themselves.

Old-school Darwinism is the most troubled theory in the history of science. Its days are numbered; our new theory of evolution will bear little resemblance to what our parents grew up believing. The treasures of a thousand Nobel prizes are buried inside the genome and the cell, waiting to be unearthed. Your part is to spread the word.

The Story Deserves to Be Told

What can you do to make this vision a reality? Nature abhors a vacuum. As with everything, the strength lies in working to *replace* something rather than just reject it.

So, we can begin to publicly argue *for* Evolution 2.0, instead of merely railing against the evolution of the past. Become a friend of the smart cell. Like the preteen Melanie in this book's preface, lots of folks will find the programming department in each cell "extremely cool" and worth knowing more about.

Take action:

- Tell those you know—in real life and in social media—about the Evolution 2.0 Prize. Maybe you or someone you know will make this discovery.
- Follow online magazines and blogs that write about evolution; counter the pseudoscience of "randomness" with links to facts and research. Feel free to cite this book and link to it. Please be cordial and always stick to the facts.
- Challenge people every single time they make derogatory remarks about nature and belittle its creations. Until someone can actually build a human eye, they're in no position to assert that it's a "terrible design."
- Insist on empirical science. Not "happy chemical accidents."
- Make your case by citing the experts. I have included references on my website for every chapter in this book, as well as additional links and resources. All the ammo you need to make the case against randomness is there.
- Bring in the reformers. Remember—Bill Gates was an outsider to the computer industry. Fred Smith of FedEx was an outsider to the shipping industry. Evolution 2.0 and its successors will be dramatically shaped by outsiders in the 21st century. New models will draw insights from a wide spectrum of disciplines—mathematics, physics, engineering, art, music, and the social sciences. Evolution belongs to everyone, not just a self-appointed good ol' boys club. If you are skilled at what you do, you may have contributions to make to the field.

- More important, Evolution 2.0 can make contributions to *your* field. When you hit an obstacle, ask, "What would nature do?"
- Challenge authors, bloggers, reporters, teachers, professors, and researchers to *prove* their assertions about evolution and evolution/origin of life. You can cite the many references in the bibliography. Make the case that biology is orderly from top to bottom.
- Consider a career in genomics, bioinformatics, biosemiotics, or the Human Genome Project. You'll make more and greater discoveries now that you're armed with the knowledge that everything under that microscope happens for a reason.
- Join the conversation on my blog, CosmicFingerprints.com, and follow me on Twitter: @cfingerprints and on Facebook: www.facebook.com/2.0.evolution.
- New research confirming and extending Evolution 2.0 is coming out literally every day. Sign up to get the latest updates in your inbox at www.cosmicfingerprints.com/supplement.

You can take your next bite of food, knowing that the DNA in one hamburger or glass of milk or bowl of navy bean soup stores more gigabytes of digital information than all the data that traversed the internet yesterday.

You can wake up every morning and step out your door with a newfound awe of the information, computation, and intelligence that is all around you. It's in every blade of grass, every flower, every honeybee, every barking dog.

Life is purposeful. *Tell somebody.*

You've Reached the End of My Story. You Still Might Want to Read Further. Here's Why.

I NEEDED TO INCLUDE technical details that don't neatly fit into the story I just finished telling you. I've put them in these appendices. Most readers will find at least one of them pretty interesting; if you're a technical reader you'll appreciate them all. Here's what comes next:

- **Appendix 1 exposes the most insidious science stopper of all: randomness.** No word or concept has killed more curiosity or scientific progress than this intellectual black hole. This chapter is a guide to the use and abuse of randomness in scientific models. I offer a new, much-needed term for describing things that are neither random nor perfectly predictable. If you're a technical reader, be sure and read this.
- **Appendix 2 is a must-read if you're Jewish or Christian or wonder if faith and science can be compatible.** It takes a fresh look at evolution through the lens of Genesis chapters 1 and 2. You might be surprised at what Genesis does say—and just as important, what it doesn't. If you thought embracing faith meant throwing your brain away, or if you ever felt you had to choose between faith and science, this chapter may surprise you.
- **Appendix 3 offers a list of recommended books about evolution, covering all sides of the debate.** It includes some excellent,

not-so-well-known texts that I found far more helpful than many very popular titles.

- **Appendix 4 details exactly what you must do to win the Evolution 2.0 prize.** If you have any doubt or question about precisely what constitutes an "encoder" or a "decoder" or why DNA is a digital communication system, this appendix lays it all out in black and white. If you wonder if DNA is really a code, or if someone suggests that snowflakes are codes and you're not sure how to show they're not, this chapter will help you greatly.

APPENDIX 1

All About Randomness

He has a whole thing beyond what he's doing at that instant.
He's orchestrating.
Everything he did prior to that moment and every moment to follow are in his head—he's hearing it all—it's choreographed.
It's very clear to him where he's going…
It's "not random."

—PHOTOGRAPHER CARRIE NUTTALL, FROM HER BOOK *RHYTHM & LIGHT*, profiling her husband, rock drummer Neal Peart of Rush

MY OWN MUSICAL SWEET SPOT is an odd place where hard rock overlaps with jazz. One day I had the music cranked up, playing a rock/jazz piece that's right in my zone.

My wife walks into the room. "Will you please turn that down?"

"Oh, you don't like the distorted guitars?"

"I don't mind the guitar all that much actually. But I can hear the entire bass line in the other room and I can't stand the randomness."

"Randomness?! That's not random. It's fractal!"*

---

* Fractal patterns in music are well known. Bach is famous for this. Repetitive sequences are rampant in music and also in genomes. A new school of genomics insists "fractal genomes grow fractal organisms." See (802) and http://fractogene.com/full_genome/r_evolution.html by Andras Pellionisz.

She steps back and crosses her arms. She squints her eyes and looks at me, as if to say, *Don't you lie to me, boy.*

"Don't you hear the melody with all its cool twists and turns?"

"That music has no melody," she retorts.

I grin at her and plead my case. "I swear, this is not random. You see, in jazz the whole idea is to get as far from the melody as you possibly can, without completely detaching from it."

"If that music has a melody," she replies with a shrug, "I sure don't hear it."

To the average guy, John Coltrane sounds like random notes on a saxophone.* For almost a century, people have been watching genes move around and mutations emerge in DNA, and, like my dear wife who isn't so much into jazz, most assumed that all those adaptations were just random. But in every crowd, there are a few who can hear the notes. They pick out the patterns. They love the subtlety.

In biology, and indeed in the entire history of science, there has always been the loud majority who missed the message that nature was quietly whispering. And there's always been that small minority who sensed order where others only saw chaos.

The genome is like jazz. If playing jazz is like solving equations, so is unlocking the mysteries of our evolutionary past. And when I say solving equations, I mean exactly that. For over a century and a half, we were told that evolution is guided by nothing but blind, pitiless selection. But it's not true. Evolution 2.0 is guided by fantastic mathematics that we are only beginning to understand.

In this appendix I'm going to show that randomness, as defined and employed in Darwinism, is antiscientific.

Half the thesis of this book is that randomness does not create codes; and that once they exist, randomness can only destroy them. The other half of this thesis is that the origin of life required the creation of codes, and that nonrandom, linguistic adaptations of DNA continue to create codes and thus drive biological evolution (645).

---

* It took me a long time to really "get" jazz. Not easy, but rewarding. I have noticed that a disproportionate number of top people in many fields, including science, business, and marketing, love jazz—especially jazz improvisation. I don't think this is an accident. Their appreciation for the abstract and unpredictable is part of what makes them great innovators in their day jobs. I also find a jazz lover is likewise less likely to dismiss evolutionary changes as "random events."

The word *random* gets tossed around an awful lot, and many times nobody's sure exactly what it means. It would be very helpful at this point to define and describe randomness in full detail.

Randomness Inflation

As mechanisms like Horizontal Gene Transfer and Transposition have started to become household words in biology, the meaning of the word *randomness* in biology has morphed. Jerry Coyne's book *Why Evolution Is True* puts it this way: "The term 'random' here has a specific meaning that is often misunderstood, even by biologists. What this means [in context of evolution] is that *mutations occur regardless of whether they would be useful to the individual*" (105, italics Coyne's).

Many biologists agree with Coyne's statement. To many, random means "non–goal-seeking," but they do not mean to say mutations don't obey the rules of known processes like Transposition. In biology, the word *random* has been used sloppily and loosely.

A large part of the entire profession has adopted a definition of randomness that is dramatically different from what engineers, physicists, and mathematicians mean by the same word. Sloppy language fosters sloppy thinking and all manner of misunderstandings, especially when translated to the public.

By redefining random to mean "non-teleological" instead of what random actually means (I cite a rigorous definition in the next section), one manages to escape the increasingly obvious fact that many theorists erred on a major point about the behavior of evolution: Mutations themselves happen in an orderly way.

This redefinition is as though someone told you, "The letters in this magazine are arranged randomly." Then they said, "Wait, I didn't really mean that. Actually it's the sentences that are arranged randomly." Then they changed their mind again: "No, the pages are arranged randomly." Then they say, "Okay, none of the magazine is actually random. What I meant to say is that the magazine doesn't have a goal." So, in other words, they're just pushing randomness from the outside of the Russian doll to the inside... from the bottom layers to the top. It's randomness inflation.

By redefining randomness, they push crucial questions into the shadows. It's been a way of avoiding the elephant in the room—the fact that, contrary to popular belief, evolution hasn't gotten rid of God or settled the God question *at all*. On the contrary, it suggests a creator that's even more amazing than we previously thought.

In chapter 19 we looked at mounting evidence that the mutations are also goal directed (645, 664). Research continues to accumulate that indicates cells adapt based on data collected from their environment. If mutations are goal directed, they're not random—neither in the traditional sense nor in the sense used by many of today's biologists.

Onward to a proper discussion of randomness, based on its formal mathematical definitions.

Definition of Randomness

The *Oxford English Dictionary* (2nd ed.) defines *random* in this way: "Having no definite aim or purpose; not sent or guided in a particular direction; made, done, occurring, etc., without method or conscious choice; haphazard."

More importantly, in statistics, *randomness* is defined as "governed by or involving equal chances for each of the actual or hypothetical members of a population; produced or obtained by such a process, and therefore unpredictable in detail."*

Likewise, statistics formally defines a *random process* as a repetitive one whose outcomes follow no describable deterministic pattern, but rather exhibit a probability distribution, so the relative probability of each outcome can be calculated. For example, when you roll a fair six-sided die in neutral conditions, you say it's random because before the die is rolled, you don't know what number will show up. However, the probability of rolling any one of the six numbers can be calculated if each is equally likely.

In information systems, randomness is non-order or non-coherence in a sequence of symbols, such that there is no intelligible pattern or combination. Formally, *a string of numbers or letters is random if and only if it cannot be generated by a formula that's shorter than the string itself* (800). Thus, as soon as you announce a pattern is random, you

* *Philosophy of Statistics* (801) offers a number of definitions of randomness on page 35, including this one and the one in the paragraph below.

have accepted that **no further parsing or analysis of that pattern is possible.**

Claude Shannon, the towering scientist who founded information theory in 1948, defined information entropy as the degree of uncertainty of a transmitted message, caused by noise. Noise is the addition of randomness to a signal due to extraneous factors. The most common sources of noise in electrical circuits are heat collisions of electrons and radio interference from the sun. They have the exact same effect on radios and televisions that radiation has on DNA: They destroy information.

Why Does the Neo-Darwinist Theory of Randomness Fail?

An excellent reference for randomness is *Information Randomness & Incompleteness*, by the renowned mathematician Gregory Chaitin. In the very first paper, "Randomness and Mathematical Proof," Chaitin says:

> Although randomness can be precisely defined and can even be measured, a given number cannot be proved to be random. This enigma establishes a limit to what is possible in mathematics.
>
> Almost everyone has an intuitive notion of what a random number is. For example, consider these two series of binary digits:
>
> 01010101010101010101
> 01101100110111100010
>
> The first is obviously constructed according to a simple rule; it consists of the number 01 repeated 10 times... Inspection of the second series of digits yields no such comprehensive patterns. There is no obvious rule governing the formation of the number, and there is no rational way to guess the succeeding digits. The arrangement seems haphazard; in other words, the sequence appears to be a random assortment of 0's and 1's.
>
> The second series of binary digits was generated by flipping a coin 20 times and writing a 1 if the outcome was heads and a 0 if it was tails. (800)

When Darwinism says evolutionary changes are accomplished through random mutations and natural selection, this potentially means the three following things at the very least:

1. Before, during, or after DNA replication, any particular letter (e.g., "GAC") could be randomly changed to any other letter ("CAG"), and the change might occasionally confer a benefit to the organism. This is the definition of a random mutation.
2. When DNA is copied, portions of the DNA strand might be accidentally folded or reversed, causing entire groups of letters to be miscopied. The change might occasionally confer a benefit to the organism.
3. Traditional Darwinism emphatically denies that evolutionary mutations in DNA are goal seeking, directed, or obey any specific pattern. Neo-Darwinists emphatically assert "the essential Darwinian notion of 'spontaneous,' 'accidental,' or 'chance' variation with respect to adaptation" (128, 129, 106). Randomness is at the heart of the Neo-Darwinian explanation, and as we've learned more and more about the genome, Neo-Darwinists have had to morph the definition of randomness to maintain their position.

The Neo-Darwinian Modern Synthesis also denies Lamarckism, the idea that acquired knowledge or traits are passed on to offspring through some kind of mutation process, because that would not be random. Darwinism does acknowledge that some kinds of mutations are more common than others, and some parts of the genome are more subject to mutations than others (129).

The problem with randomness is that it always has to be defined within a specific frame of reference; otherwise, scientific precision gets lost. Randomness is always *with respect to something*. When you roll dice (which we all naturally think of as random), you don't know the outcome because you do not have rigid control of how the dice fall. But in the absolute sense, the number that comes up is not random, because how the dice bounce and land is, after all, precisely determined by the laws of physics.

The reason dice are random *to you and me* is because our hands are too imprecise to control the outcome. The imprecision of our hands when we roll those dice is our frame of reference. A raindrop falling on your windshield is random *with respect to the motion of your car.* It is not random with respect to the cloud it fell from.

When biologists speak of random mutations in DNA, they are espousing a theory that the cell is not in control of the changes, in the exact same way that you are not in control of the dice when you play Monopoly. They are saying that the changes to DNA come from outside the system and the cell's ability to control what happens. We've seen throughout this book that this is not the case.

Across all of physics there is a larger question of whether anything is truly random at all. Depending on your interpretation of quantum mechanics, it might be possible that everything is determined by laws, and the only limitations are our ability to observe those laws. This is controversial and invokes questions far beyond the scope of this book. But in general, randomness is always a last resort, because scientific investigation stops as soon as you invoke it.

"I Have Heard Our Digital Future, and It Is Awful"

So said one of my highbrow stereo magazines when the first CDs came out. I was 14, and for two years the closest I could come to CD players was reading about them. Most people were praising CDs for their crystal-clear sound and utterly silent background. But a tiny band of die-hard audiophiles hated them. According to them, CDs sounded "harsh," "clinical," and "sterile," especially compared to their beloved vinyl records. LPs had a warm glow like a table lamp; CDs were like the blinding light in a dentist's chair.

CDs slice music into a million tiny pieces and reconstruct them. This is utterly unlike a vibrating piano string or a record groove, which is smooth and continuous. To most folks, vinyl records meant ticks and pops and needles skating mercilessly across fragile grooves. But on a high-end stereo system, those same records sounded lush and natural and warmly human.

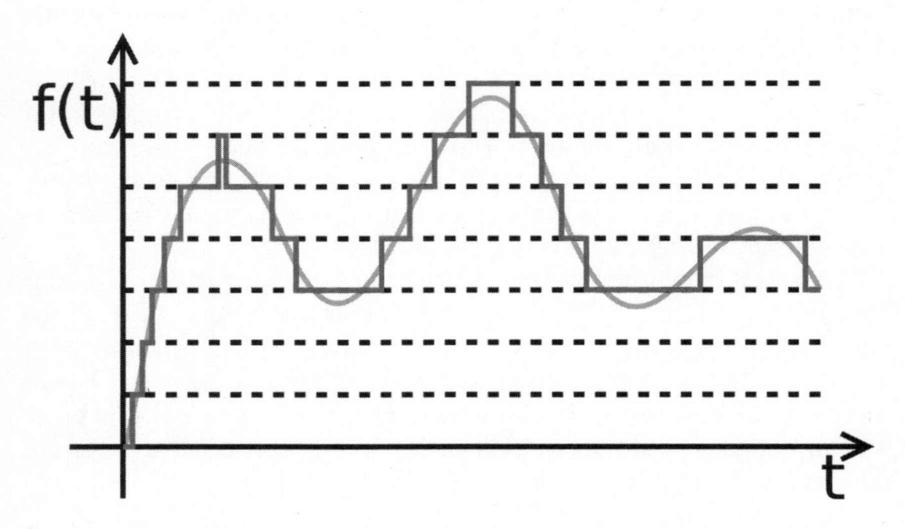

Digital recording approximates a smooth analog signal with hard digital steps. Your ears find this irritating.

One reason why CDs sounded so clinical and sterile was, if you listen to some early CDs and crank up the volume as a song fades out, you hear the music suddenly saturate with harsh distortion before it disappears entirely. You hear this because at low volume levels, the CD runs out of resolution. At that point, you only have one lonely bit representing the signal. It's either "on" or "off." The high volume level allows you to hear that fleeting rise in distortion.

The sound of that last bit switching is grating to your ears. At the tail end of a fade-out, it's 100 percent distortion. Audiophiles pointed out that this distortion was momentarily present every time the signal crossed the "zero" line, thousands of times per second. It was an artifact of converting warm, smooth analog signals into cold, hard numbers.

Dither: When Engineers Add Noise to Signals

Engineers found a way around this, called *dither* (712). Dither is a way to soften digital "glare."

Dither sprinkles noise into the signal, so it never stays at zero for very long. Instead of flipping between "1" and "0" and making an irritating clicky sound, it hovers at a noisy 0.5. The hiss that makes that 0.5 average is far less irritating to your ears. Dither can reduce the "grain" of digital images. It neutralizes rounding errors.

You can see how dither makes the picture of this cat look much more natural. The cat photo on the left has no dither—notice how the artificial shading effect is the visual equivalent of audio distortion. The photo on the right contains dither.

The left-hand picture of a cat contains no dither. The right-hand picture of the same cat has dither, which makes the rounding errors—the lines in the left-hand picture—disappear by randomizing them. It softens the glare and makes the picture much more natural and pleasant to look at.

Why does this matter when it comes to Darwin's interpretation of evolution? Because traditional Darwinism claims that new evolutionary adaptations come from random mutations, or noise (136), and that natural selection is an all-powerful force that extracts useful adaptations from that noise (111).

Dither is noise at its very best, and even then, it's only useful when it's carefully applied as a tool in the engineer's toolbox. The engineer usually generates the noise anew, and mixes it in with precision.

No engineer would ever claim that dither adds meaningful content to the signal. It's just an engineering trick that masks our ignorance of the original information. Nor would any photographer claim that the cat picture earlier came from noise.

Energy Alone Can't Reverse Information Entropy

A very common reply to this is, "Sure, entropy, the tendency toward disorder and decay, always increases in a *closed* system. But you can decrease entropy within the Earth by adding energy from the outside. In other words, you can always put the cold toast back into the toaster and heat it up again."

Isaac Asimov said, "Remove the sun, and the human brain would not have developed...And in the billions of years that it took for the human brain to develop, the increase in entropy that took place in the sun was far greater; far, far greater than the decrease that is represented by the evolution required to develop the human brain."

His reasoning is that since the sun is adding energy to the Earth, evolution can proceed on the Earth, entropy within the Earth will decrease, and the laws of thermodynamics won't be violated. This comes up often. Yet even if this greatly oversimplified statement were true, it wouldn't solve the problem because, while thermodynamic entropy and information entropy obey the same math, *energy alone does not reverse information entropy.*

Putting toast in a toaster makes it hotter, but it doesn't create information. Simply adding energy to a system does not reverse information entropy in any way. That's because energy does not create or enhance information. To quote MIT mathematician Norbert Wiener, the father of cybernetics, again, "Information is information, not matter or energy" (324).

There are other models of information, such as "Kolmogorov complexity" (800), which reduces information content to the size of the program needed to create it. None of the other information models offer a way to decrease information entropy (= increase information) without intelligence.

If some natural way exists to convert energy into code, nobody's discovered it yet.

It Can't Get Selected if It Doesn't Exist

Darwinian books have repeated the awesome power of natural selection (105, 110) and, I have to agree, most creatures do not survive. Natural selection certainly eliminates a lot of players from the game. In fact, most people are unaware of how pervasive natural selection is.

What would Earth be like without natural selection? Well... how many bags of loot would shoppers buy if they had unlimited money? Without natural selection, life would multiply without limit.

Still, natural selection can only select what *exists*. Natural selection can never create anything out of thin air. It can't make cold toast hot, and, since information entropy is irreversible, natural selection can't reverse it. Natural selection is only as good as what comes before it.

Noise doesn't add, it subtracts.

Likewise, natural selection doesn't add, it subtracts.

So, *how can the classic Darwinian model work if it's a one-way ticket to decay, degradation, and extinction? How can anything evolve if Neo-Darwinism has no workable mechanism for adding novelty?*

That's a big problem. It's only one of the many reasons that classical Darwinism is in trouble today. Information entropy overturns the widespread belief that all you need is lots of time (i.e., billions of years), and "anything could be possible."* The real truth is: garbage in, garbage out. Natural selection is not powerful enough to turn garbage into something useful.

Information entropy guarantees that time is always your enemy, not your friend. We all know how quickly our cars, toasters, computers, and watches succumb to the ravages of time. (Ever switch on a computer you bought in 1994?) Not to mention the aging of our own bodies. It became apparent to me that the ubiquitous notion that large amounts of time *help* things evolve and thrive is an urban legend. If it weren't for the powerful, driving evolutionary mechanisms in chapters 11 to 16, all time could ever buy us is extinction.

Everyone with deep experience in any "Darwinian" system well knows the limitations of natural selection. When someone asks, "Why

---

* See Daniel Dennett's points about this and Darwinism being a "Universal Acid" in chapter 3. He presumes random copying errors will produce an "abundance" of useful variations. Information entropy makes this impossible. His version of evolution is the information equivalent of a perpetual motion machine that gets faster and faster every year. Information theory thus overturns many of Dennett's conclusions.

is Starbucks so successful?" it's not enough to say, "Because a hundred other coffee chains went out of business." When someone asks, "Why did the Celtics win the playoffs?" it's not enough to say, "Because everyone else lost."

If you've ever grown a business or competed in a basketball tournament, you know that a round of eliminations all by itself doesn't create anything.*

The best natural selection can do with noisy data is select the least noisy piece of data. But the data will still be noisy and inferior. Natural selection can slow entropy down, but it is powerless to reverse it.

Why the Random Mutation Hypothesis Fails

The following are the reasons why I do not accept the random mutation theory of the Neo-Darwinian Modern Synthesis:

1. It is possible to prove that a pattern is nonrandom, but there is no mathematical procedure for proving randomness. A random mutation hypothesis can never be verified. Thus the random mutation hypothesis stands in opposition to the scientific method itself.

---

* To some readers it might seem silly that I should even need to make this point; of course natural selection by its very definition doesn't create anything. But there is still a significant contingent of scientists who continue to assert, even now, that "natural selection is the only game in town" (106). Mayr and Provine's definitive textbook *The Evolutionary Synthesis* on page 3 says, "The term Darwinism in the following discussions refers to the theory that selection is the only direction-giving factor in evolution" (102). It calls people who embrace this view "selectionists."

Their insistence on selection as the only direction-giving factor is equivalent saying that no further knowledge other than selection is necessary to explain how life evolves. This is very clever because "selectionism" has succeeded in preventing the majority of scientists from recognizing the true significance of Transposition, Symbiogenesis, and so forth for almost a century.

Britannica defines Neo-Darwinism as follows: Theory of evolution that represents a synthesis of Charles Darwin's theory in terms of natural selection and modern population genetics (*Britannica Online Encyclopedia*, s.v. "neo-Darwinian," www.britannica.com/EBchecked/topic/408652/neo-Darwinism, accessed January 13, 2015). The term was first used after 1896 to describe the theories of August Weismann (1834–1914), who asserted that his germ-plasm theory made impossible the inheritance of acquired characteristics and supported *natural selection as the only major process that would account for biological evolution.* (Emphasis mine.)

In the literature, natural selection often sounds like a magic wand (680). In everyday conversations about evolution, I encounter surprisingly many people who are under the impression that natural selection possesses some sort of creative power and that the exact nature of mutations is of minor importance.

2. A hypothesis that a particular change in DNA follows some kind of rule is inherently more scientific than a hypothesis that the change is random. Why? Because science is the discovery and classification of orderly behavior.

   The significance of this becomes clear when you watch people debate this specific question:

> *Neo-Darwinist*: "There is no purpose or reason for this mutation. It just happened. Accept it."
>
> *Post-Darwinist*: "But research shows there is a reason for this mutation. It's specifically caused by the cell's response to signals from the environment. Cells rearrange their genomes in predictable, systematic ways."
>
> *Neo-Darwinist*: "No, that can't be. Randomness alone is sufficient to explain evolution, and Darwin's theory is beautiful. You're dragging mysticism back into science. I thought we got rid of this nonsense 100-plus years ago."
>
> *Post-Darwinist*: "I just gave you a systematic, documented explanation...with empirical evidence from genome sequences and biochemistry. You can use this in your battles against Creationists if you want. I'm supporting the theory of evolution by explaining gaps in the fossil record. You're rejecting it. Why?"

I've had conversations with people who grew angry, saying, "No, Perry, evolution doesn't happen fast, it happens slow, and it doesn't follow any sort of structure. It happens by accident!" It seems they feel threatened by the idea that life is purposeful.

Randomness is a lot like saying, "God did it"—except far worse. Because at least if you believe God did it, then you can reasonably assume it functions in an orderly way. Randomness is a scientific brick wall, because it predisposes people to see mistakes and absurdities where others find order and structure.

In fact, "randomness" is wholesale abdication. Some Darwinian purists stopped looking for systematic mutation mechanisms 70 years ago, while "heretics" are still finding them. Merlin's paper (129), which I mentioned earlier in the "Why Does the Neo-Darwinist Theory of Randomness Fail?" section, insists on the non-purposeful nature of mutations, and is an attempt to dismiss the contrary evidence.

If randomness brings us to a dead end, then classical Darwinian evolution is nothing but a 3.8-billion-year string of singularity events. None of them can be further investigated beyond "The fittest survive and everything else dies!"

3. Nowhere in engineering communication theory or computer science is noise added to a signal to increase its information content. Sure, noise has limited uses—a rap musician or DJ might find a use for dither in digital signal processing, for example—but it does not enhance the original signal.
4. Theodosius Dobzhansky did extensive experiments bombarding fruit flies with radiation for decades (see chapter 4). This induced DNA mutations, which Dobzhansky expected to accelerate evolution—but it didn't. In fact, we have an abundance of evidence that random mutations destroy DNA. Direct evidence that they improve DNA is in exceedingly short supply (637).
5. Cells devote significant resources to protecting DNA from copying errors, and to correcting them when they do happen. This means that cells do not *want* random copying errors. It also shows they have the ability to recognize them when they occur.
6. We've witnessed viable systematic methods of DNA mutation. Symbiogenesis was first described in the late 1800s and it obeys rules. Transposition was discovered in the 1940s and it obeys rules. Genome Duplication was also discovered in the 1940s and also obeys rules. Epigenetics, too, was first described in the 1940s, and epigenetic changes respond to specific environmental changes. Horizontal Gene Transfer was discovered in the 1950s and it obeys rules. The very definition of a transposon (see chapter 6) implies nonrandom behavior, because a transposon is a coding sequence that mutates differently than other sequences. All of these systems can be reliably triggered in scientific experiments (645). Genetic engineers achieve desired results by using premeditated environmental shocks, manipulating cells to edit their genomes.
7. People in engineering and computer science build all kinds of models. I have a friend, Andrew, who writes computer programs that model Forex (foreign currency) trades. Andrew tells me that whenever you build a model like that, you *only* choose a random variable as an absolute last resort. Wherever possible, you prefer

to model behavior more precisely than that, with some kind of formula, correlation, or predictive algorithm. The best biologists apply the same standards to evolutionary models. We should be looking for other models before landing on randomness, and not take randomness as gospel.

8. Your immune system generates new antibody combinations to fight invaders when immune cells rearrange their DNA. They gather data by generating great variability—over 10^{11} binding specificities (661)—and then hyper-selection takes over. This is called V(D)J recombination. Immune cells read the relevant parts of their own edited DNA strand and build new antibodies. If immune cells were just mutating randomly, you'd die long before your immune cells ever found the correct antibodies. In immune cells, variability is combined with targeting. The variability is built in, not left to random events. It occurs at well-defined locations so the proteins have the right structures to function. The only reason you're alive is that every day your immune system engineers mutations that were once believed impossible (661). McClintock's plants purposefully repaired their DNA in response to damage (664); your immune system purposefully develops new DNA combinations to produce specific antibodies to fight specific invaders.

Granted, in very rare instances, copying errors can end up being useful. The name "Google" was originally supposed to be "Googol" (which means 10^{100} in math lingo) but the person who registered the name spelled it wrong. The misspelled name stuck.

Like Google's now-famous name, it is inevitable that some vanishingly small fraction of adaptations were caused by random copying errors. I freely admit that. However, it is not possible to prove those events were random. Nor is it possible to build any predictable, testable evolutionary model based on such assumptions. (I know of no other story where a copying error resulted in a famous, worldwide brand.)

You can, however, build predictable, testable models based on Symbiogenesis, Horizontal Transfer, and the other Swiss Army Knife mechanisms in this book. They are the drivers of Evolution 2.0. The positive role of randomness is minuscule and its destructive role is huge.

Lenski's Evolving Bacteria Lab

Richard Lenski is a MacArthur Fellowship prizewinner and evolutionary biologist at Michigan State University in East Lansing. I met him at a TEDx conference in Chicago where we talked at length. He's famous for a long-term evolution experiment with *E. coli* that started in 1988 (104) and has since monitored genetic changes in 12 populations of bacteria. By 2014, the populations had reached 60,000 generations.

The populations started out identical. Some adaptations were noted in all 12 populations. Others were more limited. One achievement was a new strain of *E. coli* that was able to digest citric acid even though it came from a strain that was previously unable to digest citrate.

His experiments are remarkable. I admire him for his work in this area. I disagree with his conclusions, however. In a paper he co-authored, "Historical Contingency and the Evolution of a Key Innovation in an Experimental Population of *Escherichia coli*" (104), Lenski insists mutations are random, yet the results within this very paper show bacteria making identical adaptations multiple times.

It would be impossible for anyone to prove the changes are random (800). What we can easily prove, however, is that the probability of the exact same adaptation happening twice by accident is inconceivably small. If these mutations were truly random, the same outcome could scarcely ever occur twice.

Is Natural Selection "Maxwell's Demon"?

For evolution to happen at all, *entropy must be reversed.* Adaptations must produce higher and higher levels of order. If natural selection is all you need to reverse entropy, all is well for Darwinism. If natural selection can't reverse entropy, then Neo-Darwinism is in trouble.

Fortunately, the folks in thermodynamics have given us a handy tool to picture this. It does a great job of clarifying what natural selection can and can't do. This tool is a character called "Maxwell's demon."

Maxwell's demon is a theoretical character from thermodynamics who counteracts the natural forces of entropy by opening and closing a door to move cold molecules to one side and hot molecules to the other. He's a molecular heat pump.

Maxwell's demon is a conceptual figure in physics who stands between a hot box and a cold box. Every time a hot molecule on the cold side approaches the divide between the boxes, he opens a door, allowing the hot molecule to pass through, making the hot box hotter and the cold box colder. Thus he reverses entropy. Natural selection is often claimed to do this. However, since natural selection cannot select just one bit at a time, but can only select entire genomes at once (millions or billions of bits), natural selection cannot reverse information entropy. Cells, in contrast, actively edit their genomes at the bit level. (659)

If you're standing at a window, you can behave like Maxwell's demon and get flies out of your house by letting out a single fly. But you can't do the same thing with a garage door, because the opening is too large; more flies will come in than go out. Natural selection is likewise powerless to craft complex subtle changes, because the unit of selection is not one base pair or gene, but an entire organism.

Maxwell's demon is like you standing by a window, trying to open it just in time for an annoying fly to exit your house, without letting another fly in. If flies are more plentiful outside than inside, getting the flies inside to go outside is an uphill battle.

Maxwell's demon can only work by transferring one molecule, one bit or one fly at a time. If you have to open an industrial-sized garage door instead of a tiny window, 40 flies will come in for every 10 that go out. You ain't never gonna to get rid of your flies!

In the analogy between thermodynamics and evolution, natural selection is Maxwell's demon opening and closing the window. Molecules of hot or cold air are equivalent to single bits of information in the genome (1 = hot, 0 = cold). If the demon can select for individual bits as they float near the window, then Maxwell's demon can reverse the normal increase of disorder, or entropy.

Here's the problem: Even very small genomes have nearly a million bits of information. This means the smallest unit that ever passes through any "window" is a million bits.

Natural selection can't select any unit smaller than one organism. Whatever the differences between organisms, they compete as whole individuals. In biology, the smallest number of bits that natural selection can operate on at once is about 1 million. The smallest known organisms have more than 300,000 base pairs (506), with two bits per base pair. Natural selection cannot behave like Maxwell's demon because it can't choose individual bits.

This is why random mutation and natural selection can never increase the quality of information. Natural selection can reduce the speed of information entropy but it can't *reverse* information entropy, which is what must occur for evolution to take place.

Again, the best that natural selection can ever do for information entropy is slow it down. In other words, if a population of fruit flies is experiencing bad mutations, natural selection can slow the extinction of those fruit flies by selecting the ones that didn't mutate. That's the best selection can do, because natural selection doesn't create information.

One guy asked it like this: "Can the fine-tuning of one molecule (to subtly refine the behavior of an enzyme or DNA binding protein, perhaps in only one tissue or one phase of development) really be based on the life or death of the entire organism?"

Maxwell's demon answers this question: Absolutely not. Natural selection cannot do subtle refinements down to the single bit. Unless a subtle refinement is an issue of life and death, natural selection will not effectively shape it.

The cell, however, does change individual bits and groups of bits in a goal-directed way, as Barbara McClintock demonstrated. The famous physicist Erwin Schrödinger called the ability of living things to control and regulate events "negative entropy" (656).

Blind material forces in nature do produce patterns. Together with the known laws of physics, they produce things like snowflakes, tornadoes, hurricanes, sand dunes, stalactites, rivers, and ocean waves. These patterns are the natural result of what scientists categorize as chaos and fractals. These things are well-understood and we experience them every day. Chaos theory is a refined science. But . . .

Randomness = Noise and It Always Destroys: How Randomness Violates the Rules of Any Language

To understand why noise only makes data worse, it helps to look at the *Seven-Layer Model*. I can best illustrate the Seven-Layer Model for you in terms of language. Since everyone reading this book speaks a language, I'll add another illustration of the concept. It adds a term we need to make a critical distinction.

Linguists use a simpler "Four-Layer Model" that describes all human languages (202):

4. Pragmatics
3. Semantics
2. Syntax
1. Statistics

I counted down from 4 to 1 to emphasize that the highest level in language is meaning and the lowest level is the alphabet.

Let me use the linguistic model here. The lowest layer, which contains the alphabet, is called "statistics." Here's why: Everyone who speaks English knows the letters *E* and *A* appear a lot more often than the letters *Q* and *X* and *Z*. There is a statistical probability of how often these things appear. *E* appears 12.7 percent of the time. *Z* appears 0.07 percent of the time (203). The ratios are different in other languages. This is the lowest level of a language—the stats of the alphabet.

Languages also have spelling and grammar. In linguistics, this is called *syntax*. Languages have rules that say certain things have to come before other things. For example, when you make a statement like, "She is sitting on a chair," the noun (*she*) comes before the verb (*is*). But when you ask a question like, "Is she sitting on a chair?" the verb comes before the noun.

The third layer of language is meaning or *semantics*. The message that comes out has to mean something logical and sensible. It's a request or a command that you can understand. It communicates a complete idea.

The fourth layer of language is intent, which linguists call *pragmatics*. When you say, "Please bring me a cup of coffee," your intent is for the waitress to put a hot cup of coffee on your table. There's usually a reason why you say something.

The sentence

she chair on a sitting

has statistics but its syntax is incorrect—it has no meaning and the intent is likewise unclear.

These concepts apply equally to human languages and computers. In time I confirmed they also apply to biology. DNA contains no message unless the letters are organized into instructions that follow rules of genetic spelling and grammar. As I explained in chapters 7 and 19, all major concepts in communication theory and many concepts in linguistics apply to DNA (403, 326).

Noise violates the rules of every layer. Noise has a completely different statistical profile than language. Noise corrupts syntax; it corrupts semantics, and it destroys the meaning and obscures the intent.

Consider this sentence with one obvious copying error—one mutation:

Michael wa%hed his car.

The spelling is wrong. If you don't know that "wa%hed" was supposed to be "washed" then the syntax is wrong, too, because the corrupted sentence has no verb. The sentence makes no sense because it's not complete. In black-and-white terms, that one mutation trashed our entire sentence.

In a conversation, missing a sentence here and there isn't that big of a deal. But in manufacturing—or in the development of a cell—one corrupted bit or byte can wreck everything. One missing codon dooms you to cystic fibrosis. On YouTube there's a video of what happens to the video game "Super Mario Brothers" when the program is corrupted just a tiny bit. It's called "Corrupting Super Mario Bros." (703) That's why in computers, each layer has its own separate error correction system: so errors don't accumulate and crash the program. So does DNA.

The difference between a simple code and a language is that a language is a code with multiple, separate layers:

- The rules of spelling are distinct and separate from the rules of grammar.
- The rules of grammar are distinct and separate from the meaning.
- And the most literal meaning is often separate from the actual intent.

In chapter 6 I said evolution has to strictly follow the rules of the genetic language. Let me explain what I mean.

All messages are encoded from the top down. Your intention determines what you say, which determines the sentences you form, which determine the words you choose.

When you write, the steps occurs in this order:

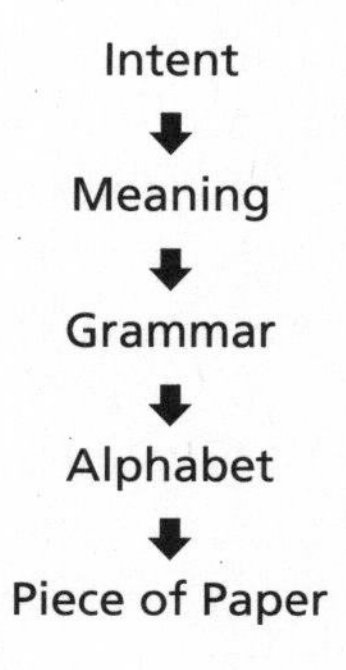

And all messages are decoded from the bottom up. You read letters on this page, which form words, which make messages that communicate my intent.

When you read:

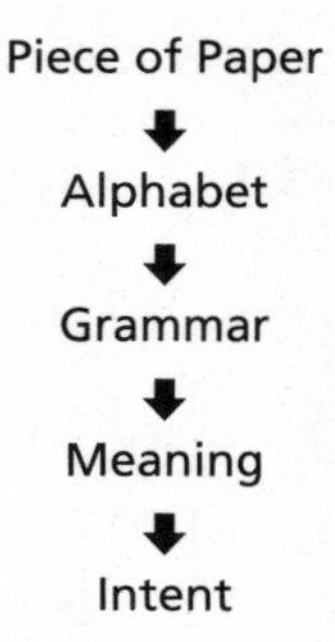

There's those Russian dolls again. You have to pack and unpack them in exact sequence. If you don't, you can't read a book or newspaper. If the cell doesn't read these layers in the genome, you get birth defects at best, or an aborted cell.

Language has layers of rules: allowable letters and numbers, rules of spelling, rules of grammar and punctuation. As I described in chapter 19, DNA is three-fourths similar to human language. Any evolutionary process has to rearrange genes and chromosomes according to the rules of the genetic language.

It's Not Random. It's Not Deterministic. It's *Ergodic*.

Random means when a system changes, it follows no particular pattern. *Deterministic* means that given two identical situations, the system will do the exact same thing each time because it's utterly predictable. But when cells change their DNA, their response is neither random nor deterministic. It's actually somewhere in between.

Claude Shannon chose an obscure word to describe this in-between place: *ergodic*. It's a crucial concept in his 1948 paper "A Mathematical Theory of Communication" (320), because ergodic patterns come in very handy when you need to detect and correct errors.

If you look up "ergodic" in a math book or Wikipedia, you'll get a highly abstract explanation that's not very helpful to most people. It essentially means "irregular regularity." In plain English, an example of ergodic behavior is what I mentioned about *E* appearing 12.7 percent of the time and *Z* appearing 0.07 percent of the time. This is so predictable that you can tell English from French or German from Spanish just by counting letters. That's because the letter frequencies are distinctly different in each language. *Z* is 17 times more common in German than it is in English, but it's even less common in Danish (0.03 percent).

Ergodic also describes certain letter patterns appearing in very predictable amounts, as in "*i* before *e* except after *c*." The most common word in English is the word *the,* and the hundredth most common word is *us* (914). It also means you'll reliably see this same pattern over and over whether you're reading Charles Dickens, the Bible, or Stephen King. But another aspect of ergodic behavior is that even though the word frequencies are extremely predictable, and sentences obey grammatical rules, *the same sentence almost* never *occurs twice.* If you turn to page 5 of any book and find the fifth sentence on the page, odds are that exact sentence has never appeared anywhere in any other book in history.

I just turned to page 5 of my edition of Charles Dickens' *A Tale of Two Cities*. The fifth sentence says, "The emphatic horse, cut short by the whip in a most decided negative, made a decided scramble for it, and the three other horses followed suit." If you search for that exact phrase on Google or Amazon, you'll confirm: sure enough, nobody besides Charles Dickens ever wrote that exact sentence with those same 26 words in a row.

Likewise, if you go find the fifth sentence of the fifth email you sent this morning, whatever you said in that sentence was also probably unique; so far as you can tell, neither you nor anyone else ever wrote that exact same sentence.

As speakers of English, we know why this is so. It's because even though we use a common dictionary and well-known grammatical rules, our response to a specific person is tailored to the situation. There are so many words and possible combinations of words that just stringing the same 26 words together twice is highly improbable.

The concept of ergodicity delineates a major distinction between two types of improbable events. The probability that you were born where you were born, to the parents you were born to, and that you ended up being right where you are right now at this moment in your life, is vanishingly small. But every one of those improbable events follows amazingly predictable, ergodic patterns. The chances of your parents having a person exactly like *you* were incredibly small. But the chances of your parents having a child of your skin and eye color and height and weight were very high.

You don't know exactly what questions will appear on the SAT, but you still know what *kinds* of questions you'll find on the test. You don't know what Carlos Santana is going to play on his next album, but the particular sound of his guitar will be unmistakable. Santana's signature sound is a form of ergodic behavior. It just operates at the pragmatics level of language instead of the statistical level.

English is ergodic, live jazz improvisation is ergodic, and cellular behavior is, too. It is impossible to subject two identical cells to a situation that is utterly identical in every microscopic aspect. And although we do sometimes find that cells engineer identical solutions to the same problem (104), their response does not appear to be purely "mechanical" or deterministic. It strikes me, as a communications engineer, that their response can best be described as *linguistic*. Ergodic patterns are par for the course in all linguistic structures.

The pattern that all vertebrates have a similar overall skeletal structure is ergodic. The pattern that animals, regardless of species, have very similar organs—lungs or gills, stomach, liver, heart, brain, eyes, and ears—is ergodic.

Because most people lack a term like *ergodic*, they lapse into black-and-white thinking, assuming that if it's not exactly predictable, it must be random instead. But the next DNA rearrangement is no more "random" than the next email you send to a friend. That rearrangement occurs *in context*.

Dr. Jean-Claude Perez, who spent decades as an information scientist at IBM, wrote an utterly fascinating paper, reporting a precise ergodic pattern in the genome, called "Codon Populations in Single-Stranded Whole Human Genome DNA Are Fractal and Fine-Tuned by the Golden Ratio 1.618" (316). He shows that the percentages of 64 different codons in a single chromosome follow a fractal stair-step pattern that is based on the famous Golden Ratio, 1.618..., from mathematics.

ONLINE SUPPLEMENT

The Golden Ratio and the mathematics of DNA

www.cosmicfingerprints.com/supplement

The Golden Ratio is the famous number seen in the architecture of the pyramids, the Parthenon, and the human body. The Greeks, Leonardo da Vinci, and countless artists and architects have embraced this ratio in designs over the last 2,500 years.

This precise mathematical stair-step pattern of letter frequencies in DNA is strikingly different from anything we find in human language. It suggests that the linguistic rules of the genome may be highly mathematical in a way that English is not. Perez calls this pattern a *checksum*. Checksums are well known to communication engineers and programmers; they are ubiquitous in networking and data storage. After each message comes in, the receiver runs a calculation to make sure the message isn't corrupted. This is a little bit like counting playing cards to make sure you have a complete deck: regardless of the order of the cards, you make sure your deck has four aces, four queens, four kings, and so forth.

Perez speculates that this mathematical structure could even be the key to understanding how Barbara McClintock's "jumping genes" (transposons) knew where to jump. This is pure conjecture, but my guess is that the genome is like a 100-dimensional sudoku puzzle or Rubik's Cube, where a certain number of squares have patterns that are allowable and others that are not. When faced with threats, the organism furiously rearranges the matrix, making its best guess as to what's going to work best.

Sometimes the organism succeeds. That's how it evolves. If the truth is anywhere close to this, then one of the most exciting projects of genetics in the 21st century will be cracking the code of the genetic Rubik's Cube.

Riches Hidden in Secret Places

King Tut's tomb was lost for millennia, until Howard Carter, persuaded that rumors of the ancient tomb were true, embarked on a mission to search for it in 1907. But his financier, Lord Carnarvon, grew frustrated with Carter's lack of progress, and in 1922 informed him he would fund one last season.

Fortunately, some stone steps were located during their digging, and on November 26, 1922, Carter chiseled away a corner of a doorway, and by candlelight he could see many gold and ebony treasures still in place in what would become known as the most intact tomb in Egypt's Valley of the Kings. He struck literal gold, and today Tut's treasures are among the most traveled of any ancient exhibit.

This is precisely where we are with DNA. Now that the random mutation and junk-DNA theories have been debunked, and knowing that Evolution 2.0 obeys the rules of a fascinating mathematical matrix, we can now begin to explore the treasures inside. An exciting century of discovery lies ahead of us.

APPENDIX 2

Genesis 2.0

We are stardust, billion-year-old carbon,
We are golden, caught in the devil's bargain,
And we've got to get ourselves
Back to the garden

—JONI MITCHELL

RECENTLY I HAD LUNCH WITH PAUL, a Christian who is president of a high-tech company in Chicago. In about 10 minutes it was obvious he could converse intelligently on any topic. He had a degree in chemical engineering, and as we began to explore evolution, he impressed me with his grasp of a wide range of subjects, from business to astronomy.

That day, we were talking about Creation, evolution, and the Bible.

Paul said to me, "My conviction is that the Bible teaches a young Earth. I believe the Earth is 6,000 years old. I take this position because I feel it is necessary for me to be intellectually honest as a Christian." He leaned back in his chair and continued, with a perplexed look on his

ONLINE SUPPLEMENT

Young Earth vs. old Earth: Does YEC simply "interpret the data differently"?

www.cosmicfingerprints.com/supplement

face. "But Perry, I will readily admit to you that I cannot defend that with empirical science; I've never been able to see any way to work it out."

His friend Jeff* was sitting next to him. Jeff is the founder of an affluent consulting firm in Chicago. He is also a Christian, but does not have a science background. Jeff said, "God could make the universe in a billion years, or he could make it in an instant. I don't think it matters."

I said to Jeff, "I appreciate what you're saying, and in theory that might be true. But if the universe was created 6,000 years ago, yet has the *appearance* of having an exquisitely detailed 13-billion-year history, that drags us into all kinds of freakish philosophical problems."

Paul nodded. He knew exactly what I meant. Some folks are untroubled by such questions, but people in the sciences are never content to shrug off such things. I admired Paul for his candor about his Young Earth Creationism. After all, he was right: It is not scientifically defensible. It runs into trouble with the speed of light and radiometric dating, plus astronomy, geology, archaeology, and a half dozen other disciplines.

In efforts to defend a young Earth, Young Earth Creationists have embraced many theories. My brother Bryan and I were both initially discouraged by this. (In fact, "Young Earth" was the first pair of rose-colored glasses that shattered for Bryan, long before he dug into deeper questions about evolution.) Paul, also a chemical engineer and technologist, freely acknowledged the problems. He said, "I don't like talking about this with non-Christians, because I don't like to lead with my chin."

Ouch.

One of my friends from high school was less demure:

> If what I believe goes completely against what a sinful world considers "highly unpopular," then I know without a doubt that I am on the right track. Evolution is a theory created by Satan himself to blind the eyes of the unbelieving.

I wonder how her son's faith is going to hold up when he grows up and finds out he can prove a star is a million light years away, alone at night with his telescope... and he doesn't need to take any unbelieving, sinful scientist's word for it. I think she's making a big mistake by adopting such a rigid interpretation of Genesis.

---

* Paul's and Jeff's names have been changed.

Many Christians reject major tenets of modern science because they feel faith prohibits them from accepting an old Earth or evolution. Whether you agree or disagree with them, their willingness to suffer scorn and ridicule for reasons of conscience does deserve respect.

Still, science does not testify to a young Earth. Painfully, many people feel forced to choose between faith and science. But I don't believe it is necessary for a religious person to compromise faith *or* science.

Putting Faith on the Science Chopping Block

In chapter 3 I described how my faith was wavering. I said to myself, *I'm going to let science and engineering answer this question for me*. That story gives some conservative Christians a heart attack. They say, "You can't do that!"

Well, I did. And right or wrong, like it or not, *lots* of people do it. Many walk away from faith because it contradicts what they consider to be plainly obvious facts, including my brother Bryan and many of your own friends.

You may ask, "Perry, why did you come to a different conclusion than your brother and so many others?" I believe the answer is, *because I demanded real-world results*. You'll recall that in chapter 3 I said, "Just as I had gotten to the bottom of physics in speaker design, I knew I could reduce evolution to a set of core principles." I *knew* what it felt like to understand a subject from the bottom up.

I believe the main difference between science that leads people *to* faith and science that leads people *away* from faith is that real science is *empirical*. When Bryan claimed that millions of random mutations over millions of years would inevitably *have* to lead to the hand at the end of his arm, he was making a big assumption. He was making an imaginative leap, because nobody had ever demonstrated this was true.

Much of Darwinism consists of just-so stories of warm ponds accidentally producing life, and millions of copying errors accidentally producing precisely the adaptation that was needed at exactly the right time. Atheism assumes an entire fine-tuned universe can pop into existence for no reason at all.

None of this is empirical. We are bullied to believe it simply because it gets repeated many times...along with dubious claims like "virtually all scientists accept this." Consensus is not science.

Empiricism showed me that all codes whose origins we know have been designed. Empiricism said evolving codes always obey layers of linguistic rules. Experience showed evolution is always teleological. So when an atheist tells you life self-organized near hot vents in the ocean, demand proof. When he tells you natural selection is the Universal Acid that removes all need for designers, ask him to use blind natural selection to debug software.

Coming full circle, I realized that the Genesis story can be held up to observation, at least to a degree. It states a particular vantage point and a specific sequence of events over time. You can check it against the geological record. This is why people today are still challenging the Genesis story when creation myths from other traditions have long been ignored.

As I was doing my research, I also began hunting for a way of reading Genesis that does match what we know of Earth's history. There are many, many interpretations of these passages, and when one didn't seem to match our current knowledge of science, I set it aside and kept looking. The view that follows is a work in progress, but it seems to work well so far.

Matching Genesis to Empirical Science

I prefer the reading of Genesis 1 and 2 that follows, because it matches modern cosmology, geology, and the fossil record nicely. In this chapter, I'm going to share with you what I said to my friend Paul, the chemical engineer turned high-tech company president.

As we read Genesis together, let's make two assumptions:

1. The writer is describing events *as they appear from the surface of the Earth* starting with verse 2, which establishes the point of view for the remainder of the chapter.
2. "Day" is a period of time, not 24 hours. The Hebrew word for day (*yom*) has a variety of meanings in Genesis. A day can be a moment, an era, or a thousand or even a billion years (949). In Genesis 2:4, for example, the word *day* is used to refer to the entire Creation sequence!*

* The New American Standard Bible, known for translating the original text as literally as possible, renders Genesis 2:4 as follows: "This is the account of the heavens and the earth when they were created, in the *day* that the LORD God made earth and heaven."

GENESIS (New International Version)	SCIENCE INTERPRETATION
Chapter 1	
1 In the beginning God created the heavens and the earth.	The text literally means, "At the beginning of time, God (who already existed) created everything out of nothing." Today we understand that the Big Bang was the beginning of matter, energy, space, and even time itself, all expanding from a single point in a very precise manner. The Big Bang theory was resisted for decades in part because of its resemblance to Genesis 1 and its metaphysical implications.
2 Now the earth was formless and empty, darkness was over the surface of the deep, and the Spirit of God was hovering over the waters.	This verse establishes the physical vantage point that is used from here forward. Four billion years ago, following the Hadean period, the Earth was a hostile, stormy, turbulent, water-covered ball. The Hebrew word for "hovering" is also used elsewhere in Genesis to describe an eagle protecting her young in the nest. The specific indication of God's presence in physical space seems significant. Science tells us that the earliest life forms began in the ocean 3.5 billion years ago.

GENESIS (New International Version)	SCIENCE INTERPRETATION
3 And God said, "Let there be light," and there was light. 4 God saw that the light was good, and he separated the light from the darkness. 5 God called the light "day," and the darkness he called "night." And there was evening, and there was morning—the first day.	The atmosphere changes from dark to cloudy. Light can now shine through Earth's thick cloud cover. Now there is day and night on the surface of the Earth. The phrase "There was evening, and there was morning" is a traditional Jewish expression of completion, rather than a literal evening and a literal morning.
6 And God said, "Let there be an expanse between the waters to separate water from water." 7 So God made the expanse and separated the water under the expanse from the water above it. And it was so. 8 God called the expanse "sky." And there was evening, and there was morning—the second day.	The water cycle begins. Clouds condense and form the ocean. Water evaporates from the ocean and forms clouds.
9 And God said, "Let the water under the sky be gathered to one place, and let dry ground appear." And it was so. 10 God called the dry ground "land," and the gathered waters he called "seas." And God saw that it was good.	The continents ("Pangaea") rise above the surface of the ocean, forming land and sea.

GENESIS (New International Version)	SCIENCE INTERPRETATION
11 Then God said, "Let the land produce vegetation: seed-bearing plants and trees on the land that bear fruit with seed in it, according to their various kinds." And it was so. 12 The land produced vegetation: plants bearing seed according to their kinds and trees bearing fruit with seed in it according to their kinds. And God saw that it was good. 13 And there was evening, and there was morning—the third day.	Plants appear before animals. Notice that the wording in the Bible says the "land produced vegetation." This does not rule out an Evolution 2.0 process. Also notice that it talks about plants bearing seeds and fruit according to their various kinds. It takes much longer than 24 hours for any of these things to happen.
14 And God said, "Let there be lights in the expanse of the sky to separate the day from the night, and let them serve as signs to mark seasons and days and years, 15 and let them be lights in the expanse of the sky to give light on the earth." And it was so. 16 God made two great lights—the greater light to govern the day and the lesser light to govern the night. He also made the stars. 17 God set them in the expanse of the sky to give light on the earth, 18 to govern the day and the night, and to separate light from darkness. And God saw that it was good. 19 And there was evening, and there was morning—the fourth day.	Up to this point the atmosphere has been thick and opaque. It is not possible to see the sun and moon as distinct objects in the sky. The atmosphere clears. Now sun and moon are visible. The moon and stars were already casting their light on day 1 (see verse 3), but were not visible as distinct objects until day 4. This detail is important! Otherwise the story does not make scientific sense. Our modern knowledge of the solar system predisposes us to read the Creation story from an "outer space" vantage point, but it is written from the same point of view ancient people would have read it in—from the Earth. Notice that 24-hour days are not even mentioned before day 4.

GENESIS (New International Version)	SCIENCE INTERPRETATION
20 And God said, "Let the water teem with living creatures, and let birds fly above the earth across the expanse of the sky." 21 So God created the great creatures of the sea and every living and moving thing with which the water teems, according to their kinds, and every winged bird according to its kind. And God saw that it was good. 22 God blessed them and said, "Be fruitful and increase in number and fill the water in the seas, and let the birds increase on the earth." 23 And there was evening, and there was morning—the fifth day.	The Earth is dominated by birds and fish. Insects and dinosaurs are also present on the Earth during this time but the author is not concerned with them. "According to their kinds" is scientifically correct. Dogs never give birth to anything but dogs. However, we know experimentally that Hybridization and Symbiogenesis produce species 1 + species 2 = species 3 in short periods of time.
24 And God said, "Let the land produce living creatures according to their kinds: livestock, creatures that move along the ground, and wild animals, each according to its kind." And it was so. 25 God made the wild animals according to their kinds, the livestock according to their kinds, and all the creatures that move along the ground according to their kinds. And God saw that it was good.	Earth is dominated by large mammals. Ancient Jewish people would have thought in terms of livestock.

GENESIS (New International Version)	SCIENCE INTERPRETATION
26 Then God said, "Let us make man in our image, in our likeness, and let them rule over the fish of the sea and the birds of the air, over the livestock, over all the earth, and over all the creatures that move along the ground."	God refers to himself as "us" (Elohim)—God is plural.
27 So God created man in his own image, in the image of God he created him; male and female he created them.	Man is a spiritual being, the first creature made in God's image. Unlike the animals, man is both body and spirit. This is why the origin of man's body plan, which is adapted from lower animals, doesn't alter his spirit identity as a child of God.
28 God blessed them and said to them, "Be fruitful and increase in number; fill the earth and subdue it. Rule over the fish of the sea and the birds of the air and over every living creature that moves on the ground."	Man is given the gift of caring for the Earth and the environment.
29 Then God said, "I give you every seed-bearing plant on the face of the whole earth and every tree that has fruit with seed in it. They will be yours for food. 30 And to all the beasts of the earth and all the birds of the air and all the creatures that move on the ground—everything that has the breath of life in it—I give every green plant for food." And it was so.	Verses 29–30 are often taken to imply that all creatures were vegetarians before the Fall. But that would also imply that animals and people had nothing to eat until verse 29. This view seems excessively literal. I see it as a proclamation of abundance and of the sufficiency of the Earth to nourish itself.

GENESIS (New International Version)	SCIENCE INTERPRETATION
31 God saw all that he had made, and it was very good. And there was evening, and there was morning—the sixth day.	The text says all was very good. It doesn't say it was paradise or perfection.
Chapter 2	
1 Thus the heavens and the earth were completed in all their vast array.	God ceases from his creative work on the seventh day.
2 By the seventh day God had finished the work he had been doing; so on the seventh day he rested from all his work.	Notice there is no statement, "And there was evening, and there was morning—the seventh day."
3 And God blessed the seventh day and made it holy, because on it he rested from all the work of creating that he had done.	We are living in the seventh day now.

If we had to make obtuse, complex assumptions in order to make this fit, we'd have a big problem. But our assumptions are few, simple, even elegant. This poetic 3,500-year-old text matches modern science remarkably well.

As little as 100 years ago, the prevailing scientific view conflicted with Genesis 1. The "steady state" theory of the universe—that a stable universe existed eternally into the past and future—was in vogue. But Genesis produced a testable hypothesis. Since then, astronomy, geology, and biology have shown Genesis was right and the science of the time was wrong.

I am unaware of any other ancient text, scripture, or religious tradition whose creation story comes close to Genesis in its accuracy. The Jewish scriptures compete admirably in the 21st century.

I do not get the impression that the writers of the Hebrew Bible were particularly concerned with the science questions I raise in this book. The story is much more interested in our relationship to God and our human condition.* Genesis conveys far more layers of meaning than

* Old Testament scholar John Walton puts it this way in his book *The Lost World of Genesis 1*:

I've even begun to unpack here. Yet, you can't help but ask: How did a bunch of desert nomads in 1500 B.C. get this sequence of events right? I submit to you it's because God told his prophets what He did.

In this book I've made a case that living things have a Designer. I've shown, with Evolution 2.0, that evolution is not random or accidental, but an engineered process. The remarkable accuracy of Genesis serves as evidence for me to be a Christian and not a deist.

St. Augustine said, "God wrote two books: The Bible and the book of Nature." Modern science also informs our interpretation of Genesis. Each tells us about the other. My supposition is that it's necessary to use science to help us interpret the Bible—precisely because Romans 1 says the cosmos tells us truths about God, as I described in chapter 1.

The Bible cannot be interpreted in a vacuum. Just as you can't understand Jesus' parables without knowing a thing or two about mustard seeds and agriculture, you can't understand the true meaning of Genesis without help from science. Science is not our enemy. It's our friend.

Young Earth Creationism, the Speed of Light, and the Age of the Earth

If a star in the sky is a million light years away, how old is the light from that star that reaches your eyes today?

Light travels 186,282 miles (299,792,458 meters) per second.

Some stars in the sky are a million light years away.

Therefore the light from those stars is a million years old.

Therefore the universe cannot be 6,000 years old.

As far as I'm concerned, that's the end of the conversation about the age of the universe. Simply based on the speed of light, there is not even a remote possibility that the universe is young.

"If someone came into a play late and asked the woman in the seat next to him how the play began, he would not expect her to reply with information concerning the construction of the set, the playwright's experiences, or the casting of the players. This would not be an incorrect answer. It would simply be addressing something other than what was being asked. We must be careful not to make that same mistake with the Bible. If we superimpose this illustration on the cosmos, taking it as the play, science might be seen as addressing the questions about the set, the cast, and the script. Scripture is more interested in the play itself, even though God, Producer/Director, is the link between the two and was involved on both ends." (946)

There are three kinds of responses I've seen Young Earth Creationists make to this: (1) The speed of light is changing; (2) God made a universe 6,000 years ago that looks much older than that; or (3) time is subject to Einstein's laws of relativity, and Genesis is reporting time from a different point of reference.

Regarding #1, if the speed of light were changing even slightly, the internet would seize like a blown car engine. Precise relationships in physics and engineering that mesh perfectly like gears in a watch would instantly unravel. Even things as seemingly trivial as the maximum length of a USB cable are based on the speed of light. As an electrical engineer, I have a special appreciation for all the facets of physics and engineering that cleanly fit together and hinge on the speed of light.

Even Einstein's beloved $E = mc^2$ that equates matter and energy would crumble like burnt toast, because c is the speed of light. If it's not a constant, even conservation of matter and energy goes out the window. If c isn't a constant, nothing works. Only a physicist can fully appreciate the mess this would make.

Dear reader, I hope I can convey how precise many things in physics and astronomy are. Creationists often complain about the inexact nature of carbon dating, for example. But the speed of light is not like carbon dating. It's precise to nine decimal places! Physics is for the most part an exact science. Distances and velocities of stars are exact.

NASA sending the Voyager probes into interstellar space and slinging them around planets for gravity acceleration is exact. When a physicist tells you a star is 10 million light years away, you can take that number to the bank. And if you have a telescope in your backyard, you can figure the star's distance out for yourself.*

As for #2, saying that God made a universe with an exquisitely detailed *illusion* of history that never actually happened makes it impossible to be certain of anything. You could just as easily say God made all of us six days ago with our memories intact. It means that when God

* When I began this journey, I understood this precision well from engineering. There's no debate among physicists about the speed of light. Such facts are clear and uncontroversial, especially in contrast with ambiguities that theologians debate each other about, like biblical interpretations and the meaning of the word *day*. So I decided, if the Bible really does mean to say the earth is 6,000 years old...then it's not a reliable document. I think biblical writers were unconcerned with the age of the earth. Young Earth Creationists have saddled Genesis with an unnecessary burden of literalness and missed its poetry. I understand that they simply hold biblical authority above scientific authority, and I credit them for maintaining their convictions. I am not challenging biblical authority. I am challenging their interpretation.

made the universe, he engineered 12.999 billion years of fake historical evidence.*

Such a position invokes terrible theological and epistemological problems. It makes science itself the study of illusion. It forces you to say theology and science can never line up because the universe itself is lying to us. The universe is telling the truth. And the author of Genesis is not a liar, but a poet.

And #3 just affirms that one cannot afford to be too rigid in one's interpretation of the word "day."

Literal Days?

The most common argument for a 24-hour day is Exodus 31:16–17: "The Israelites are to observe the Sabbath…for in six days the Lord made the heavens and the earth, and on the seventh day he rested and was refreshed" (901).

However, the command in Exodus is not about the length of days, it's about the pattern of work and rest. Notice there is no statement, "And there was evening and morning, a seventh day" like there is for the other six. We're in the seventh day now. This is confirmed in Hebrews 4:3–4:

> Now we who have believed enter that rest, just as God has said,
> "So I declared on oath in my anger,
> 'They shall never enter my rest.'"

And yet his works have been finished since the creation of the world. For somewhere he has spoken about the seventh day in these words: "On the seventh day God rested from all his works." (901)

---

* Occasionally I've even heard friends suggest that God made the universe this way to test people, so as to force them to choose between science and the Bible. I couldn't disagree more. Judaism and Christianity are *historical* religions. The truth claims of the Bible are rooted in historical events—Israelites escaping Egypt, the Jewish kings, archaeology, New Testament names and places. Biblical history is remarkably accurate. When Jesus asked, "Which is easier: to say, 'Your sins are forgiven,' or to say, 'Get up and walk'?" (Luke 5:23 [901]), he was confirming his spiritual authority with physical proof. In other words, his theological claims were not just heady philosophical theories. He reinforced them with physical demonstration.

Last, I have a hard time envisioning all the events of Genesis 2 (Adam naming all the animals and more) fitting into one day as we define the term. A 24-hour interpretation is a force-fit.

I invite you to listen to Hugh Ross' lecture "New Scientific Evidence for the Existence of God" (www.cosmicfingerprints.com/hugh-ross-origin-of-the-universe), where an astrophysicist explores the history of Big Bang science and its relation to Genesis in much greater detail. Physicist David Snoke also wrote an excellent book, *A Biblical Case for an Old Earth*. If you feel that the Bible advocates a young Earth, I strongly encourage you to read Dr. Snoke's book.

Was There Pain, Suffering, and Death Before the Fall?

American evangelicals widely believe the early Earth was perfect, there was no death, and Adam and Eve introduced physical death to the world. This belief is based on Romans 5:12, which says,

> Therefore, just as sin entered the world through one man, and death through sin, and in this way death came to all people, because all sinned...

Read the rest of Romans 5 with care. Ask yourself, "What kind of death is Paul talking about?" Notice the symmetry of life and death in verses 17 and 18 (emphasis mine):

> For if, by the trespass of the one man, **death** reigned through that one man, how much more will those who receive God's abundant provision of grace and of the gift of righteousness reign in **life** through the one man, Jesus Christ! Consequently, just as one trespass resulted in condemnation for all people, so also one righteous act resulted in justification and **life** for all people.

Death is defeated by life. Is he saying *physical* death reigned through Adam and *physical* life came through Christ? No, he is speaking spiritually in both cases. He is not talking about physical death. He says this affected *people* (no mention of plants or animals).

Genesis never says there was no physical death; it never indicates the world was perfect, or that it was a heavenly paradise. (It was also staffed by a very clever serpent from the word go, was it not?) The only reference to immortality anywhere in the Genesis story is the Tree of Life. Genesis only says it was good,* and that the Garden of Eden was a very special place for a season.

If you stop and picture it, our planet Earth with no death makes no sense. How could Adam and Eve eat even the allowed fruit without killing the cells it's made of? Wouldn't that be death?† Is it conceivable for elephants to never ever step on an ant? If the Earth was once perfect, it wasn't an Earth that science has any familiarity with at all. In that case, those who embrace Young Earth Creation should not attempt to reconcile their views with science. (There are some who do not.)

If your Sunday school teacher told you the early Earth was a perfect paradise and that picture got fixed in your mind, it's hard to get rid of. You probably didn't think about it very hard at the time. You certainly didn't know that for thousands of years many theologians have taken a more nuanced view.

Evolution 2.0 even clarified, for me, Saint Paul's words in Galatians 5:17–18 (901): "For the flesh desires what is contrary to the Spirit, and the Spirit what is contrary to the flesh. They are in conflict with each other, so that you are not to do whatever you want. But if you are led by the Spirit, you are not under the law."

Everybody on Earth knows what this battle between spirit and flesh is like. What got us here (the development of our physical bodies) won't get us where we want to go (a harmonious society), even if the vehicle were the most extraordinary natural genetic engineering you can imagine. That's because natural selection still wars against equality, love, and human rights. Human evolution needs to take on a brand new meaning, one that is spirit led.

In my conversations with Young Earth Creationists, the hardest pill for them to swallow is not the idea that some of the elements in the Genesis story aren't strictly literal. It's the idea that the Earth could have been wild, hostile, or in any sense "Darwinian" at the outset. Some cannot

---

* My late colleague Michael Marshall asked, "Which is more dangerous? (1) A world with pathogens like viruses and bacteria, or (2) a world where the second most powerful being in the universe is a vicious killer who is boiling with envy, hanging around waiting for an opportunity to devour someone? Nevertheless God declared the world to be very good, despite the fact that peril was built in to the picture before man ever showed up."

† Some claim that there was plant death before the fall, but no animal death. The Bible makes no such a statement, and there is no intrinsic difference between the death of a plant cell and an animal cell.

conceive that a good God would make Earth that way. But nowhere does the Torah or New Testament ever say the Earth was perfect.

Young Earth and Eschatology

Often when I suggest to Young Earth Creationists that Earth was never a perfect paradise, the person I'm speaking with feels as though their view of the future has become unhinged. That's because a common assumption built into the Young Earth view is that future paradise is a return to the past paradise; that the whole arc of God's relationship to humanity is about restoring former perfection. Everything was once perfect; death and sin entered the world; death and sin will someday be conquered; and perfection will return.

Isaiah 11:6 says,

> The wolf will live with the lamb,
>   the leopard will lie down with the goat,
> the calf and the lion and the yearling together;
>   and a little child will lead them.

Many Christians believe that the lion used to "lie down with the lamb" before the Fall, but I don't find support for this in the Bible.

Some Protestant Christians see the Tree of Life reappear in Revelation 22 and because of this, assume Earth was once perfect. They presume Revelation describes a return to perfection that once existed before. This view is especially prevalent among American evangelicals. Nothing else in the Bible supports this. Genesis offers us no detail about the Tree of Life. There is no indication that Adam and Eve ever benefited from it. It only represents a possibility of eternal life; it plays no active role.

Other streams of Christianity, like Catholic and Eastern Orthodox, do not bring these same "perfect Earth" assumptions to the table.

The issue of a "very good" but imperfect Earth causes people a lot of heartache. Allow me to explain.

Atheists and Young Earth Creationists Share Something in Common

They share profound difficulty with the idea that a perfect God would make a world where suffering is built into the equation from the word go.

Atheists are quick to point out that any sort of "theodicy" (God's goodness in the presence of evil) immediately demolishes the possibility of an all-knowing, all-loving God.

Creationists reject evolution for very similar reasons: They believe God would never make a world that includes suffering and declare it "very good." Often they express revulsion at this idea. The book of Job deals with the pain-and-suffering question head-on. Again, the picture of ancient Earth as a death-free paradise does not come from the Bible. It is imposed from the outside. That's why it's bad theology, and it's partly to blame for the present war between science and religion.* To non-Christian people, I apologize for Christians who at times have misrepresented or ignored science because they distrust it.

ONLINE SUPPLEMENT

A theology of evolution

www.cosmicfingerprints.com/supplement

"In the Beginning Was... Information"

This book shows why origin of information is absolutely central to the secret of life and the cosmos. How interesting it is, then, to note the opening verses of John:

> In the beginning was the Word, and the Word was with God, and the Word was God. He was with God in the beginning. Through

* This view of the earth before the Fall also relieves human beings of guilt for the physical death we see around us.

For instance: Dad takes Billy for a walk through the country and they see the bloody carcass of a rabbit devoured by a wolf. "Dad, why did a wild animal kill the rabbit?" Billy asks.

"Because when Adam and Eve sinned, death entered the world. That rabbit died because of your sin and my sin."

Dad just saddled Billy with guilt for something Billy had *nothing* to do with. We may be responsible for our spiritual life and death, but physical death? Not our burden to carry.

> him all things were made; without him nothing was made that has been made.

Word = information. I don't think it's a coincidence that John anticipated the gravity of the Origin of Information question 2,000 years ago, or that this question is just as relevant today as the day it was written.

Scholars Have Written About Old Earth for Millennia

Some people insist that prior to modern science, pretty much everyone believed the Earth was 6,000 years old (923). Young Earth Creationist Ken Ham has famously said Christians have only recently compromised their reading of Genesis to accommodate modern science since the 1800s (923). This is demonstrably false.

The prominent church father Origen of Alexandria (A.D. 184–253) wrote in his book *On First Principles*, "To what person of intelligence, I ask, will the account seem logically consistent that says there was a 'first day' and a 'second' and 'third,' in which also 'evening' and 'morning' are named, without a sun, without a moon, and without stars, and even in the case of the first day without a heaven?" He said, "Surely, I think no one doubts that these statements are made by Scripture in the form of a type by which they point to certain mysteries" (918).

Origen is not discrediting Genesis. He is rather saying that the most literal, childlike reading doesn't wash. He's saying Genesis, like all of Scripture, employs symbolism, metaphor, and subtlety. We must read it as literary adults.*

Justin Martyr (A.D. 103–165), a major figure in the early Church, said, "For as Adam was told that in the day he ate of the tree he would die, we know that he did not complete a thousand years. We have perceived, moreover, that the expression, 'The day of the Lord is as a thousand years,' is connected with this subject" ("Dialog with Trypho the Jew," chapter 81, A.D. 155 [950]).

This view is not limited to a handful of early Church scholars. Jewish sources long before modern science speak of a very old Earth. In the

---

* The Swiss theologian Karl Barth used the word *saga* to describe scripture passages that are not myth and not legend, but that also are not bare historical facts. Saga refers to real historical events, which are expressed in poetic terms because human language is incapable of adequately expressing acts of God.

book *Immortality, Resurrection, and the Age of the Universe: A Kabbalistic View* (930), the authors review Talmudic writings and ancient traditions concerning the age of the Earth:

> The Midrash states that, "God created universes and destroyed them." One of the important classical Kabbalistic works, *Ma'arekhet Elokut*, states explicitly... that the Midrashic teaching that "there were orders of time before this [creation]" is also speaking of earlier Sabbatical cycles.
>
> A Talmudic passage seems to support this view of Sabbatical cycles. According to the Talmud, and some Midrashim as well, there were 974 generations before Adam. The number is derived from the verse "Remember forever His covenant, a word He commanded for a thousand generations" (Psalms 105:8). This would indicate that the Torah was destined to be given after one thousand generations. Since Moses was the twenty-sixth generation after Adam, there must have been 974 generations before Adam. The *Ma'arekhet Elokut* states explicitly that these generations existed in the Sabbatical cycles before Adam's creation.
>
> The concept of pre-Adamic cycles was well known among the Rishonim (early authorities), and is cited in such sources as Bahya, Recanati, Ziyyoni, and *Sefer ha-Hinnukh*. It is also alluded to in the *Kuzari*, and in the commentaries of the Ramban and Ibn Ezra...
>
> Rabbi Isaac of Akko writes that since the Sabbatical cycles existed before Adam, their chronology must be measured, not in human years, but in divine years. Thus, the *Sefer ha-Temunah* is speaking of divine years when it states that the world is forty-two thousand years old. This has some startling consequences, for according to many Midrashic sources, a divine day is 1,000 earthly years long, and a divine year, consisting of 365¼ days, is equal to 365,250 earthly years.
>
> Thus, according to Rabbi Isaac of Akko, the universe would be 42,000 × 365,250 years old. This comes out to be 15,340,500,000 years, a highly significant figure. From calculations based on the expanding universe and other cosmological observations, modern science has concluded that the Big Bang occurred approximately

> 15 billion years ago. But here we see the same figure presented in a Torah source written more than seven hundred years ago!
>
> I am sure that many will find this highly controversial. However, it is important to know that this opinion exists in our classical literature...
>
> Although human beings may have existed before Adam, he was the first to acquire a special spiritual sensitivity and be able to commune with God*...
>
> As this discussion demonstrates, classical Torah sources not only maintain that the universe is billions of years old, but present the exact figure proposed by modern science. There are two accounts of Creation in the Torah, the first speaking of the spiritual infrastructure of the universe, which was completed in seven days. This took place some 15 billion years ago, before the Big Bang. The second account speaks of the creation of Adam, which took place less than six thousand years ago.
>
> What is most important is that there is no real conflict between Torah and science on this most crucial issue. If anything, Torah teachings are vindicated by modern scientific discoveries.

I quote these passages not because I embrace Kabbalah or even because I'm proposing these interpretations are correct. I mention this because it proves there was much discussion about an old Earth, with a lengthy process of life's development, long before the dawn of modern science.

Despite this, some people believe if they don't guilt or frighten people into adopting their extremely literal reading of Genesis, the whole of Scripture will unravel and Christianity will fall into the abyss, taking the world with it.

Really? What a pile of fear-based, paranoid baloney. No other book has remained more current with the shifts and tumults of civilization than the Bible; we've been refreshing our understanding of scripture for 3,000 years. Why stop now?

---

* Ideas about pre-Adam humans go back to the early church. Many writers spanning almost 2,000 years, including Origen and Gregory of Nyssa, have speculated that Adam was not a single historical person. David Livingstone's book *Adam's Ancestors: Race, Religion, and the Politics of Human Origins* exemplifies this view, exploring many interpretations across history. This view is taken by many modern Jewish and Christian scholars like John Walton. I believe this view has strong merits.

If you are a mature adult who has studied the Bible since childhood, you understand certain things clearly that you found puzzling when you were young. This is because your accumulated life experience sheds light on it. We all understand human nature better at age 50 than at age 20. Science is an important facet of life experience.

Modern science speaks much more definitively about the age of the Earth than the Bible does, so it is perfectly legitimate to allow science to settle this question.

. . . Which Brings Us to the Second Creation Account in Genesis

Genesis 2 describes the formation of Adam and Eve and speaks to the essential nature of man. Every interpretation of this text is based on assumptions, whether they are stated explicitly or not. These are the assumptions I'm using as I approach Genesis 2:

1. The "breath" God breathes into Adam is not oxygen. It is spirit. It refers to man being made in God's image.
2. "Rib" is an English word for a Hebrew word that elsewhere in Genesis is interpreted "side" or "chamber." It doesn't literally mean rib.
3. "Flesh" refers to the whole human being, including will and emotions; not merely muscle and bone. Flesh is body and soul.

With these assumptions, let's look at Genesis 2 from a 21st-century perspective:

GENESIS (New International Version)	SCIENCE INTERPRETATION
7 Then the LORD God formed a man from the dust of the ground and breathed into his nostrils the breath of life, and the man became a living being.	Humans and animals alike are made of dirt. Here, God takes a body and grants it a spirit. The Hebrew word for "living being" according to BlueLetterBible.org means "soul, self, life, creature, person, appetite, mind, living being, desire, emotion, passion." This is why we see a cultural "Cambrian explosion" in ancient history. Art, writing, language, worship, architecture, agriculture, and religious ceremony suddenly appear in a geological instant.
But for Adam no suitable helper was found. 21 So the LORD God caused the man to fall into a deep sleep; and while he was sleeping, he took one of the man's ribs and then closed up the place with flesh. 22 Then the LORD God made a woman from the rib he had taken out of the man, and he brought her to the man.	Rather than literally meaning "rib," I think it means something more than merely physical. God transformed a male creature into the first human being. The man did not know he was alone—God did. Here, God takes something from the man and gives it to the woman, making her his partner—and leaving him incomplete.
23 The man said, "This is now bone of my bones and flesh of my flesh; she shall be called 'woman,' for she was taken out of man."	Being made in the image of God changes not only the way we see ourselves, but our relationship to others.
24 That is why a man leaves his father and mother and is united to his wife, and they become one flesh.	If you go to any Jewish or Christian marriage conference, they'll invariably say that "becoming one flesh" is far more than a physical act of sex. It's emotional and spiritual oneness. Sex in a loving relationship is a deeply spiritual experience. Why? Because the "rib" in verse 21 refers to something that is also emotional and spiritual. Only in a loving marriage is this brokenness reconnected. This is why Adam felt complete with his new wife, and it's why a dead marriage is the worst form of loneliness.

As I mentioned earlier, Talmudic sources suggest that generations existed before Adam and Eve, but they weren't spiritual beings—they weren't truly human.

Genesis genealogies are sometimes presented as evidence for a young Earth. Bishop James Ussher famously used them to declare Sunday, October 23, 4004 B.C. as the day the Earth was created (945). However, cross comparisons of genealogies throughout the Bible show that most of them "telescope," or skip generations. Genesis 11 skips Cainan, whom Luke places between Shelad and Arphaxad. We do not know for sure whether Genesis 5 skips generations, but if it doesn't, it's the exception, not the rule.

In any case, what's remarkable is not that we can somehow manage to reinterpret the Genesis story so it fits an old Earth and evolutionary history. What's remarkable is that renowned rabbis and theologians explicitly held similar views many centuries before modern telescopes and cosmology. As Solomon said, "There is nothing new under the sun."

Who's Your Daddy? Is Your Daddy a Chimpanzee?

Imagine your kidneys are failing. Then just when everything appears hopeless, you learn that the Mayo Clinic has successfully transplanted a kidney from a chimpanzee to a human. If you accept that same organ transplant and save your life, does that make you any less human?

I think not. The essence of a human being is not our body. It's our spirit. It's the fact that we are made in the image of God. We do not physically look like God, but we reflect God's spiritual essence.

Don't forget that Christians believe God became man, physically born of a human mother. "The Word became flesh and dwelt among us," as St. John wrote. If a human can be the Son of God by possessing the Spirit of God, then why can't a primate become a human being by receiving a human spirit?

If being born of a woman doesn't denigrate God, then having the same genes as a hominid doesn't denigrate man—or prevent man's spirit from being made in the image of God.

No Christian should be offended by common ancestry, because who we truly are—our real identity—doesn't come from our flesh in the first place. It comes from spirit.

It's Not Necessary to Be Dogmatic About Interpretation

Here I have presented my best current interpretation of Genesis 1, Genesis 2, and modern science. It's not written on stone tablets. I offer this to you provisionally. Our interpretations can and will evolve as discoveries are made. I've not even begun to explore all the different views, but so far, this is the one that makes the most sense to me.

ONLINE SUPPLEMENT

Christian college prof: "What I say when my students struggle with evolution"

www.cosmicfingerprints.com/supplement

You yourself are not bound to one single interpretation that you're obligated to defend for the rest of your life. Scientists and theologians alike both understand that our models of reality are a work in progress.

The Judeo-Christian story continues to age well. The word *gene* comes from the same Latin root as the word *Genesis*. Despite thousands of years of opposition and bitter assault, no other document has aged as gracefully as the Torah and the story in Genesis. In the 21st century, genes and Genesis are alive and well.

Evolution Speaks of a Very Capable Designer

Evolution "should be" impossible. It turns disorder into order, a complete reversal of normal entropy and decay. It's still beyond the reach of man to create. Compared to a single cell, man's greatest achievements are puny, insignificant. Yet "impossible" evolution is happening all around us, in the lab and in the wild.

The lethargy of Darwinism was assuming that evolution happens by sheer accident. In these pages we've seen why it's impossible for anything to evolve by pure accident. The scientific understanding of Evolution 2.0 is that a magnificent genetic program guides it as it drives forward. To incorporate this into your worldview demands a greater estimation of God than Creationists were ever willing to allow.

Darwinists Underestimate Nature. Creationists Underestimate God.

I believe in Evolution 2.0 because the God I believe in is more magnificent than previously believed. He doesn't have to beam zebras from the sky onto the savanna. He designed a process that formed them from the dust of the ground and tailored them to their environment.

When we accept this view of God, it cracks the door to study evolution and gain insights into the process God used. God wants us to study *all* of what He has made—not just part of it. God is the Original Scientist, the Original Engineer. This opens huge vistas in medicine, genetics, computer science, and technology. You can't learn how zebras are built from a miracle—but you can learn from a natural process.

Come, sit at God's feet, and learn. The smorgasbord of knowledge lies before you. "It is the glory of God to conceal a matter; to search out a matter is the glory of kings" (Proverbs 25:2).

What if we understood God to be an engineer so skilled that he endows cells with the ability to engineer themselves—to form cooperative networks so a trillion cells work together, fueling the creation of a single organism? Trillions of such organisms drive the progress of an entire planet.

Some Creationists are saying, *Don't listen to those scientists. Living things can't turn into new species.*

Some Darwinists will say, *Don't believe some silly book about engineered evolution. DNA doesn't have "goals" and cells don't "engineer" themselves. Natural selection and lots of time are the only things you need.*

Between these narrow views lies a vast unexplored territory, where everything we learn about cells and nature potentially empowers us to feed the hungry, to live in greater comfort, and to ponder the wonders of a truly astounding universe.

Is your God small—or is your God big?

APPENDIX 3

Recommended Books

If there's a book of Jubilations
We'll have to write it for ourselves
So come and lie beside me darling
And let's write it while we still got time

—JOSH RITTER

The Case for Common Descent

The following list includes both secular and religious books. All do an excellent job of showing that all members of the animal kingdom have a great deal in common, and that common lineage is directly inferred by many forms of scientific evidence.

Your Inner Fish by Neil Shubin
A journey into the 3.5-billion-year history of the human body by a paleontologist who sees reflections of many different creatures in the organs of human beings.

Creation or Evolution: Do We Have to Choose? by Denis Alexander
Alexander is director of the Faraday Institute for Science and Religion at St. Edmund's College, Cambridge University. He makes a positive case for evolution within the context of Christian belief.

He's unique in his ability to draw good from both sides of the debate.

The Language of God by Francis Collins

Collins is a former atheist and director of the U.S. Human Genome project, and is now director of the National Institutes of Health. In his book, he explains why he sees evidence for a Creator within evolution itself. Criticism: Collins makes the same randomness assumption that Dawkins does, but sees the hand of God in it. He'd do better to recognize the actual mechanisms of evolution, and their systematic behavior. Very little is said about evolution's ingenious systems.

Finding Darwin's God by Kenneth R. Miller

Miller is a practicing Catholic and a biology professor at Brown University. His book discusses writings by Augustine and various interpretations of Genesis. He says, "Even the God of Genesis is a Deity fully consistent with what we know of the scientific reality of the modern world."

The Selfless Gene by Charles Foster

A penetrating look into the hostility between Christianity and evolutionary science, from a gentleman who teaches at Oxford University. Not only does Foster think the war between faith and science is unnecessary; he also tells why evolution raises theological questions that Christians ought to have the courage to answer.

Books Offering Secular Critiques of Evolution

The following books are completely nonreligious and offer detailed criticisms of Darwinism, ranging from calls for reform to scathing rebuke. I've arranged them roughly in order, from least to most critical.

Acquiring Genomes: A Theory of the Origins of Species by Lynn Margulis and Dorion Sagan

The authors are Darwinists but not Neo-Darwinists. They lambast the "randomness" mutation theory and extreme overemphasis on natural selection. They present solid evidence for

Margulis' beautiful theory of Symbiogenesis. Margulis was a true pioneer in our modern understanding of evolution.

What Darwin Got Wrong by Jerry Fodor and Massimo Piattelli-Palmarini

The authors (a philosophy professor and a biologist, respectively) are evolutionists and "card-carrying atheists" who nevertheless point out that Darwin misunderstood many things and most of the problems in his theory have yet to be solved.

The Altenberg 16: An Exposé of the Evolution Industry by Suzan Mazur

This book explores a closed-door conference in which major biologists seek to formulate a new theory of evolution. Mazur, a freelance journalist, interviews each in turn, revealing a very fragmented, contentious inner picture of present evolutionary theory. She also considers financial interests and recipients of government grants that keep science in gridlock.

Evolution under the Microscope: A Scientific Critique of the Theory of Evolution by David Swift

This book surveys the incredible wonders of nature that are revealed by modern biochemistry and the electron microscope. It shows extreme levels of organization on every scale and examines many obstacles for which classical evolutionary theory fails to account.

Evolution, Old and New by Samuel Butler

This book is a perfect illustration of G. K. Chesterton's statement, "History is old things happening to new people," because it was written in 1882. The discussion is highly informed and you'll see from reading it that the shortcomings of Darwinism were as well established 100 years ago as they are now.

Mathematics of Evolution by Sir Fred Hoyle

The famous astronomer Fred Hoyle was a fervent atheist, so much so that he loathed the Big Bang theory and its implications of a "prime mover." Hoyle coined the term *Big Bang* itself, intending it as an insult. He was also a highly esteemed mathematician. He shows that the claims of Darwinism are mathematically flat-out impossible. He advocates panspermia, the view that life originated in outer space.

The Great Evolution Mystery by Gordon Rattray Taylor
Taylor is a journalist who takes a definite positive stance that evolution itself is a fact. But his book is loaded with hundreds of examples that explicitly contradict traditional Darwinian dogma. It shows that Darwin's model leaves gaping holes and engenders far more questions than answers.

Darwinian Fairytales by David Stove
Stove was a flaming atheist, a witty philosopher of some repute from Australia. He was very funny and in this book was deadly serious. Stove gives Darwin a nasty spanking by revealing grave inconsistencies between theory and practice, and glaring contradictions between separate elements of the theory.

Books with Alternative, Non-Darwinian Models of Evolution

Evolution: A View from the 21st Century by James A. Shapiro
Shapiro is a bacterial geneticist at the University of Chicago. He describes the evolutionary mechanisms I outline in this book, and many others, in exhaustive detail. Highly technical, not for the uneducated reader. The eminent biologist Carl Woese went so far as to call it "the best book on basic modern biology I have ever seen." Superb, earns my highest recommendation.

Biocentrism by Robert Lanza
A medical doctor and stem cell pioneer discusses how quantum mechanics shows that consciousness gives rise to the material world, not the other way around. An informative and easy read, and an excellent companion to this book.

Creative Evolution: A Physicist's Resolution between Darwinism and Intelligent Design by Amit Goswami
Like Lanza, Goswami, a theoretical quantum physicist, asserts that consciousness is the driving force in the universe, not matter.

Quantum Evolution: How Physics' Weirdest Theory Explains Life's Biggest Mystery by Johnjoe McFadden

A professor of molecular genetics defines the cell as a quantum computer, and considers evolutionary adaptations as directed actions by cells.

Natural Creation or Natural Selection? A Complete New Theory of Evolution by John Davidson

Davidson's dedication sums it up: "To the great biologist, Charles Darwin, who helped free us from the influence of religious dogma. It is now time to free ourselves from the influence of Charles Darwin."

A Silent Gene Theory of Evolution: A Genuine Rival to the Theory of Evolution by Warwick Collins

Postulates that genes are turned on and off, and that junk DNA is a storehouse of evolutionary possibilities, activating genes as necessary to make wholesale changes to the organism.

The Design Matrix: A Consilience of Clues by Mike Gene

This astute observer introduces a matrix for evaluating the probability of design, then proceeds to describe evolution as a system-driven phenomenon.

Codex Biogenesis: Les 13 Codes de l'ADN [*The 13 Codes of DNA*] by Jean-Claude Perez

If you can read French and you're a math geek, you will find this an utterly fascinating study of precise numerical ("ergodic") patterns in DNA, based on the Golden Ratio, 1.618... The former IBM biomathematics and AI researcher describes the "mathematical Rubik's Cube" that cells shuffle when they need to massively adapt to massive change.

The Science of God by Gerald L. Schroeder

A physicist explores how religious belief is enhanced by an open-eyed investigation of the world, and how honest science demands humility when faced with the extraordinary richness of life's creation. He bridges the gap between ancient and modern views of origins and explores the harmony between ancient scriptures and modern science from a Jewish perspective.

Virolution by Frank Ryan
When the human genome was first sequenced, not only was it discovered to be shockingly simple (only 10 times more complex than bacteria), but this medical doctor also found in the code large fragments derived from viruses—fragments that proved vital to evolution of all organisms. It turns out that viruses contribute a sixth "blade" to the Swiss Army Knife of evolution.

APPENDIX 4

The Origin of Information: How to Solve It and Win the Evolution 2.0 Prize

Birds fly over the rainbow
Why then, oh why can't I
If happy little bluebirds fly
Beyond the rainbow
Why, oh why can't I?

—Judy Garland

THIS BOOK ISSUES A CHALLENGE: "Show an example of a code that's not designed. All you need is one."

To date no one has documented the emergence of a naturally occurring code. If such a process were found, it would be one of the most celebrated discoveries of the last 100 years. A group of private investors and I are offering a prize for this discovery, if the process that produces codes can be patented. The prize caps at $10 million. The money is an offer for a majority of the patent rights. The discoverer will still retain some ownership, so if there's a large upside, he or she will potentially earn more than just the face amount.

The value of such a discovery is incalculable, because not only would it revise our fundamental understanding of physics and biology, and very possibly help us solve the Origin of Life problem, it also would achieve something very significant technologically: It would be the world's first instance of true artificial intelligence.

As such, every technology company in the world would likely be interested in it. It's potentially worth hundreds of millions, maybe even billions of dollars. Microsoft, Google, Apple, Samsung, AT&T, automotive manufacturers—you name it—they'd be hot for it. If such a thing can be developed, I would love nothing more than to participate in commercializing and marketing it. It would be a huge boon to my own technology consulting business.

If you can solve this problem, my investors and I will pay you several million dollars and become your business advocates. If the process is not patentable—if it's "public domain"—then I simply offer you $100,000 and will publicize your discovery on CosmicFingerprints.com and in releases to the major news media.

If someone discovered a naturally occurring code, at least a third of this book might become obsolete. My design hypothesis is falsifiable. I'm willing to follow the evidence *wherever* it leads. The truth is always more valuable than whatever misconceptions we've been hanging on to.

How to Prove You've Discovered a Naturally Occurring Code and Win the Prize

Information is defined as digital communication between an encoder and a decoder, using agreed-upon symbols. To date, no one has shown an example of a naturally occurring encoding/decoding system.

The following specification defines the criteria for winning the prize by discovering a naturally occurring code:

1. The award goes to the person or group who discovers a naturally occurring code, not the person or group who merely reports it.
2. Humans can design the experiment, employing all manner of state-of-the-art laboratory equipment, creating ideal conditions, et cetera. However, the submitted system cannot be preprogrammed with any form of code whatsoever, which would be cheating.

3. Since the origin of DNA is unknown, the submitted system cannot be a direct derivative of DNA or produced by a living organism. Bee waggles, dogs barking, RNA strands, and mating calls of birds don't count. Such codes are products of animal intelligence, genetically hard-coded and/or instinctual.
4. The origin of the submitted system must be documented such that its process of origin can be observed in nature and/or duplicated in a real-world laboratory according to the scientific method.
5. The submitted system must be digital, not analog. A system that transmits vibrations from one place to another, for example, or from one form of energy or another, does not count.
6. The submitted system must have the three integral components of communication functioning together: encoder, code, decoder.
7. The message passed between encoder and decoder must be a sequence of symbols from a finite alphabet.
8. A *symbol* is a group of k bits considered as a unit. We refer to this unit as a message symbol m_i $(i = 1, 2, \ldots M)$ from a finite symbol set or alphabet. The size of the alphabet M is $M = 2^k$ where k is the number of bits in the symbol. For a binary symbol, $k = 1$, $M = 2$. For a quaternary symbol in DNA, $k = 2$, $M = 4$.
9. A *character* is a group of n symbols considered as a unit. We refer to this unit as a message character c_i $(i = 1, 2, \ldots C)$ from a finite word set or vocabulary. The maximum size of the character set C is $C = M^n$. For a standard computer byte, $M = 2$, $n = 8$, $C = 256$. For a triplet group of quaternary symbols in DNA, $M = 4$, $n = 3$, $C = 64$. (Items 8 and 9 adapted from ref. 321)
10. The submitted system must be labeled with values of both encoding table and decoding table filled out.
11. For the submitted system, it must be possible to objectively determine whether encoding and decoding have been carried out correctly. For example, when you press the "A" key on the keyboard, a letter "A" is supposed to appear on the screen and there is an observable correspondence between the two. In defining gender, a combination of X and Y chromosomes should correspond to male, while XX should correspond to female. For any given system, a procedure should exist for determining whether input correctly corresponds to output.

All non-patentable submissions, along with our evaluations of those submissions, are available in their entirety for public review at the following page: www.cosmicfingerprints.com/submissions.

Why DNA Is a Code

There is direct mathematical equivalency ("isomorphism") between Claude Shannon's 1948 communication system and DNA. The following diagram appears in Shannon's paper:

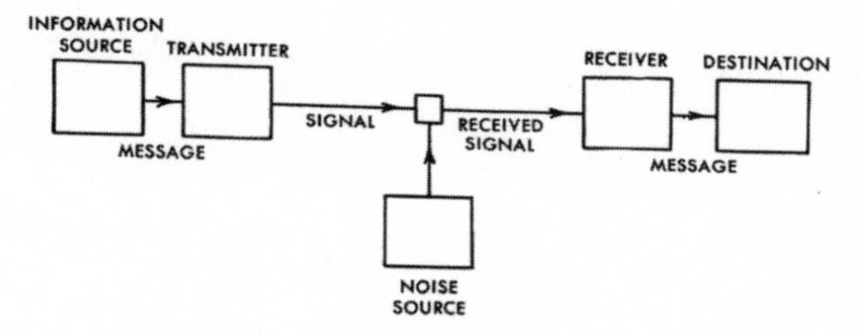

Claude Shannon's communication model; a schematic diagram of a general communication system (320).

Hubert Yockey employs Shannon's model to explain where each element of the DNA transcription / translation process fits in the scheme:

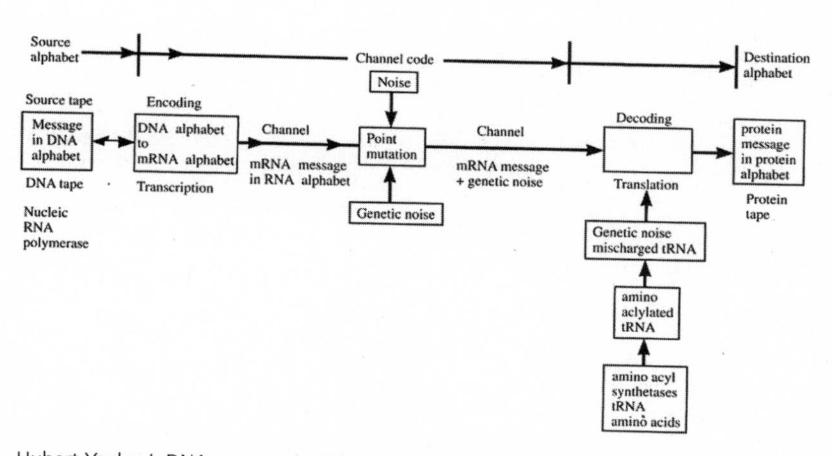

Hubert Yockey's DNA communication channel model (326). Notice that it contains the exact same components as Shannon's—the two systems are isomorphic.

Both ASCII and DNA are formal communication systems according to Shannon's model because they encode and decode messages using a system of symbols. The following examples show why DNA is not *like* a communication system, or *analogous* to a communication system; it is formally defined *as* a communication system. "Information, transcription, translation, code, redundancy, synonymous, messenger, editing, and proofreading are all appropriate terms in biology. They take their meaning from information theory (Shannon, 1948) and are not synonyms, metaphors, or analogies" (326).

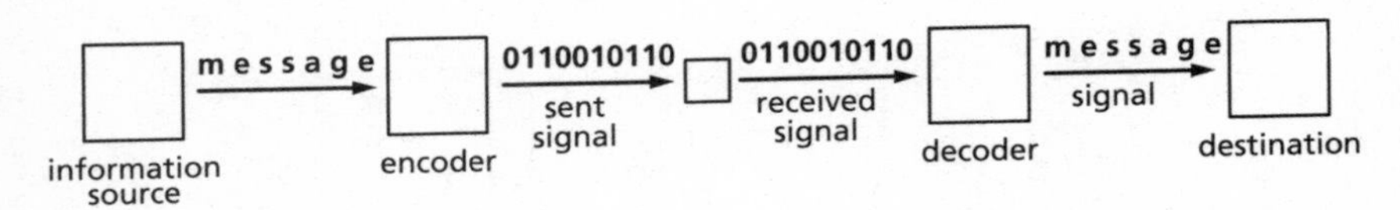

All communication systems have an encoder, which produces a message, which is translated by a decoder. A winning entry for the prize will be a natural process that generates all three of these interlocking mechanisms without anyone having to design them.

Example #1: The ASCII Code Matches Shannon's Model

Keyboard → ASCII → Computer Screen: When you press the letter "A" on the keyboard, the letter is encoded into ASCII and decoded by the computer and a letter "A" appears on the screen.

ASCII characters contain 7 symbols, so $n = 7$. The ASCII character set C is 2^7 or 128 characters.

Encoding tables for ASCII (letter on keyboard → binary code):

Input (letter on keyboard)	Encoded Message
A	1000001
B	1000010
a	1100001
b	1100010

The complete ASCII table is available in many computer science books or online (e.g., www.asciitable.com and http://en.wikipedia.org/wiki/Ascii).

Decoding tables for ASCII (binary code → letter on screen or printer):

Encoded Message	Output (displayed as an arrangement of pixels on screen or printer)
1000001	A
1000010	B
1100001	a
1100010	b

Example #2: The Genetic Code Matches Shannon's Model As Well

Nucleotides → mRNA → Proteins: Base pairs are grouped into codons and encoded (transcribed) into messenger RNA, then decoded (translated) by the ribosomes into proteins.

The DNA symbol unit is a nucleotide, forming a four-letter alphabet of adenine, cytosine, guanine, or thymine. Each base pair contains $k = 2$ bits of information. A character consists of $n = 3$ symbol units. Character set C is 4^3, which is 64 characters. DNA's redundancy scheme maps these 64 characters to 20 amino acids.

Encoding tables for DNA (base pairs → mRNA):

Nucleotides (Input)	Amino Acid (Encoded Message)
CCC	Proline
ACC	Threonine
GGG	Glycine
AAA	Lysine

The complete genetic code chart is available in most biology books (e.g., 245) and at http://en.wikipedia.org/wiki/Genetic_code#RNA_codon_table.

Decoding tables for DNA (amino acids → proteins):

Amino Acid Sequence (encoded message)*	Output Peptide/Protein† (organism name)	# of Amino Acids
YGGFM	Met-enkephalin (HS)	5
MRTGNAN	Microcin C7 (EC)	7
DRVYIHPF	Angiotensin 2 (HS)	8
CYIQNCPLG	Oxytocin (HS)	9
CYFQNCPRG	Vasopressin (HS)	9
QHWSYGLRPG	Gonadoliberin-1 (HS)	10
RPKPQQFFGLM	Substance P (HS)	11
DVPKSDQFVGLM	Kassinin (KS)	12
GGAGHVPEYFVGIGTPISFYG	Microcin J25 (EC)	21
RSCCPCYWGGCPWGQNCYPEGCSGPKV	Neurotoxin 3 (AS)	27
HSQGTFTSDYSKYLDSRRAQDFVQWLMNT	Glucagon (HS)	29
APLEPVYPGDNATPEQMAQYAADLRRYINML-TRPRY	Pancreatic hormone (HS)	36
KCNTATCATQRLANFLVHSSNNFGAILSSTN-VGSNTY	Islet amyloid polypeptide (HS)	37
CTPGSRKYDGCNWCTCSSGGA-WICTLKYCPPSSGGGLTFA	Serine protease inhibitor 3 (SG)	40
DDGLCYEGTNCGKVGKYCCSPIGKYCVCYD-SKAICNKNCT	Pollen allergen Amb t 5 (AT)	40
VGIGGGGGGGGGSCGGQGGGCGGC-SNGCSGGNGGSGGSGSHI	Microcin B17 (EC)	43
TTCCPSIVARSNFNVCRLPGTPEALCATYTG-CIIIPGATCPGDYAN	Crambin (CA)	46
ATYNGKCYKKDNICKYKAQSGKTAICKCYVK-KCPRDGAKCEFDSYKGKCYC	Antifungal protein (AG)	51
GIVEQCCTSICSLYQLENYCNFVNQHLC-GSHLVEALYLVCGERGFFYTPKT	Insulin A-B chains (HS)	51

Appendix 4: The Prize

Amino Acid Sequence (encoded message)*	Output Peptide/Protein† (organism name)	# of Amino Acids
DIPEVVVSLAWDESLAPKHPGSRKNMACY-CRIPACIAGERRYGTCIYQGRLWAFCC	Neutrophil defensin 1 (HS)	56
CSSNAKIDQLSSDVQTLNAKVDQLSNDV-NAMRSDVQAAKDDAARANQRLDNMAT-KYRK	Major outer membrane lipoprotein (EC)	58
RPDFCLEPPYTGPCKARIIRYFYNAKAGLCQT-FVYGGCRAKRNNFKSAEDCMRTCGGA	Pancreatic trypsin inhibitor (BT)	58
EEYVGLSANQCAVPAKDRVDCGYPHVTPKE-CNNRGCCFDSRIPGVPWCFKPLQEAECTF	Trefoil factor 3 (HS)	59
MDPNCSCAAGDSCTCAGSCKCK-ECKCTSCKKSCCSCCPVGCAKCAQGCICK-GASDKCSCCA	Metallothionein (HS)	61
IRCFITPDITSKDCPNGHVCYTKTWCDAFCSIR-GKRVDLGCAATCPTVKTGVDIQCCSTDNC-NPFPTRKRP	Long neurotoxin 1 (NK)	71

* For a complete legend of amino acid abbreviations, see http://www.ncbi.nlm.nih.gov/Class/MLACourse/Modules/MolBioReview/iupac_aa_abbreviations.html.
† Mature form. Source: www.uniprot.org.

This is only a partial listing of the simplest proteins. There are about a million known proteins, many of them extremely complex. More information on protein structures is available at www.uniprot.org and www.ncbi.nlm.nih.gov. Live hyperlinks for this chart are at www.naturalcode.org.

Similar tables are easily made for other codes and communication systems, like HTML, bar codes, postal codes, Morse code, computer file formats, and programming languages.

You can find the most up-to-date version of this specification at www.naturalcode.org. The online version supersedes the version in this book.

Bibliography

Citations are numbered by category as follows:

100 Darwinism & Neo-Darwinism
200 Basic Biology & Science Definitions
300 Genetic Code & Coding Theory
400 Linguistic Models of DNA
500 Cellular Cognition
600 Post-Darwinian Models of Evolution
700 Technology
800 Mathematics
900 History, Philosophy, & Theology

Pay attention to the differences between the 100s, which are generally old-school Darwinism, versus the 400s, 500s, and 600s, which are real-time, lab-tested, systems-based models of evolution. A close examination of the literature in both camps shows that major components of classical Neo-Darwinism are obsolete.

100 Darwinism & Neo-Darwinism

100 Cyber Top Cops. (2014). "Stock Market Spam—Hash Buster Text and Hash Busting." Retrieved from http://www.cybertopcops.com/stock-market-spam-hash-busting.php
101 Ashbrook, T. (2005, August 10). "On Point with Tom Ashbrook: Debate over Intelligent Design." *National Public Radio*. Retrieved from http://onpoint.wbur.org/2005/08/10/debate-over-intelligent-design
102 Ayala, F. J. (1969). "Evolution of Fitness, v. Rate of Evolution of Irradiated Populations of Drosophila." *Proceedings of the National Academy of Sciences of the United States of America*, 63, 790–793.
103 Bergman, J. (2006). "The 'Nothing in Biology Makes Sense Except in the Light of Evolution' Myth: An Empirical Study and Evaluation." Retrieved from http://www.trueorigin.org/biologymyth.php

104 Blount, Z. D., Borland, C. Z., & Lenski, R. E. (2008). "Historical Contingency and the Evolution of a Key Innovation in an Experimental Population of *Escherichia coli*." *Proceedings of the National Academy of Sciences of the United States of America, 105*, 7899–7906.
105 Coyne, J. A. (2009). *Why Evolution Is True*. New York: Viking Adult.
106 Coyne, J. A. (2012, August 22). "James Shapiro Goes after Natural Selection Again (Twice) on HuffPo." *The Huffington Post*. Retrieved from http://whyevolutionistrue.wordpress.com/2012/08/22/james-shapiro-goes-after-natural-selection-again-twice-on-huffpo/
107 Crow, J. (1972). *Proceedings of the 6th Berkeley Symposium on Mathematical Statisitics and Probability. Volume 5: Darwinian, Neo-Darwinian, and Non-Darwinian Evolution (Biostatistics of Genetical Processes in Human Populations)*. Cambridge, UK: Cambridge University Press.
108 Darwin, C. (1859). *On the Origin of Species by Means of Natural Selection, or the Preservation of Favoured Races in the Struggle for Life*. London: John Murray.
109 Darwin, C. (1888). *The Descent of Man, and Selection in Relation to Sex*. London: John Murray.
110 Dawkins, R. (1976). *The Selfish Gene*. Oxford, UK: Oxford University Press.
111 Dawkins, R. (1986). *The Blind Watchmaker: Why the Evidence of Evolution Reveals a Universe without Design*. New York: W. W. Norton.
112 Dawkins, R. (1996). *River out of Eden: A Darwinian View of Life*. New York: Basic Books.
113 Dawkins, R. (2000). *Unweaving the Rainbow: Science, Delusion and the Appetite for Wonder*. New York: Mariner Books.
114 Dawkins, R. (2009). *The Greatest Show on Earth: The Evidence for Evolution*. London: Transworld Digital.
115 Dawkins, R. (2010). *The Rise of Atheism*. Retrieved from http://www.atheistconvention.org.au/richard-dawkins/
116 Dennett, D. C. (1996). *Darwin's Dangerous Idea: Evolution and the Meanings of Life*. New York: Penguin.
117 Dietrich, M. R. (2003). "Richard Goldschmidt: Hopeful Monsters and Other 'Heresies.'" *Nature Reviews, Genetics, 4*, 68–74.
118 Dobzhansky, T. G. (1937). *Genetics and the Origin of Species*. New York: Columbia University Press.
119 Dobzhansky, T. G. (1955). *Evolution, Genetics, and Man*. New York: Wiley.
120 Goldschmidt, R. B. (1982). *The Material Basis of Evolution (Reissued)*. New Haven, CT: Yale University Press.
121 Hall, B. K. (2001). Commentary. *American Zoologist, 41*, 1049–1051.
122 Hanson, F. B., & Heys, F. (1930). "A Possible Relation Between Natural (Earth) Radiation and Gene Mutations." *Science, 71*, 43–44.
123 Hanson, F. B., & Heys, F. (1932). "Radium and Lethal Mutations in Drosophila: Further Evidence of the Proportionality Rule from a Study of the Effects of Equivalent Doses Differently Applied." *American Naturalist, 66*, 335–345.
124 Hoffer, Eric. (2006). *The Passionate State of Mind and Other Aphorisms*. Titusville, NJ: Hopewell.
125 Hoyle, F. (1999). *Mathematics of Evolution*. Memphis, TN: Acorn Enterprises.
126 Kawaguchi, H., O'Huigin, C., & Klein, J. (1992). "Evolutionary Origin of Mutations in the Primate Cytochrome P450c21 Gene." *American Journal of Human Genetics, 50*, 766–780.
127 Loony London—Men on Display in Zoo, "Just Another Primate." (2005, August 29). *LifeSiteNews.com*. Retrieved from https://www.lifesitenews.com/news/loony-london-men-on-display-in-zoo-just-another-primate
128 Mayr, E., & Provine, W. B. (1998). *The Evolutionary Synthesis: Perspectives on the Unification of Biology*. Cambridge, MA: Harvard University Press.
129 Merlin, F. (2010, September). "Evolutionary Chance Mutation: A Defense of the Modern Synthesis' Consensus View." *Philosophy & Theory in Biology*, 2. doi:10.3998/ptb.6959004.0002.003. Retrieved from http://quod.lib.umich.

edu/p/ptb/6959004.0002.003/--evolutionary-chance-mutation-a-defense-of-the-modern?rgn=main;view=fulltext

130 Meyer, S. C. (2013). *Darwin's Doubt: The Explosive Origin of Animal Life and the Case for Intelligent Design*. New York: HarperCollins.

131 Moorhead, P. S., Kaplan, M. M., & Brown, P. (1967). *Mathematical Challenges to the Neo-Darwinian Interpretation of Evolution: A Symposium Held at the Wistar Institute of Anatomy and Biology, April 25 and 26, 1966*. Philadelphia: Wistar Institute Press.

132 Morgan, T. H. (1916). *A Critique of the Theory of Evolution*. Princeton, NJ: Princeton University Press.

133 Newport, F. (2009, February 11). "On Darwin's Birthday, Only 4 in 10 Believe in Evolution." *Gallup*. Retrieved from http://www.gallup.com/poll/114544/darwin-birthday-believe-evolution.aspx

134 Sawyer, S. A., Zhang, Z., & Hartl, D. L. (2007). "Prevalence of Positive Selection among Nearly Neutral Amino Acid Replacements in Drosophila." *Proceedings of the National Academy of Sciences of the United States of America, 104*, 6504–6510.

135 Stein, R. (2012, August 9). "Gonorrhea Evades Antibiotics, Leaving Only One Drug to Treat Disease." *NPR* [Blog post]. Retrieved from http://www.npr.org/blogs/health/2012/08/10/158464908/gonorrhea-evades-antibiotics-leaving-only-one-drug-to-treat-disease

136 University of California at Berkeley. (n.d.). "Understanding Evolution for Teachers: Mutations Are Random." Retrieved from http://evolution.berkeley.edu/evosite/evo101/IIIC1aRandom.shtml

137 Yates, D. (2012, March 12). "Study of Ribosome Evolution Challenges RNA World Hypothesis" [Press release]. Retrieved from http://news.illinois.edu/news/12/0312ribosome_GustavoCaetano-Anolles.html

200 Basic Biology & Science Definitions

200 BookRags. (2013). "Dissection—Research Article from *World of Health*." Retrieved from http://www.bookrags.com/research/dissection-woh

201 Ayala, F. J., & Rzhetskydagger, A. (1998, January 20). "Origin of the Metazoan Phyla: Molecular Clocks Confirm Paleontological Estimates." *Proceedings of the National Academy of Sciences of the United States of America, 95*(2): 606–611. doi:10.1073/pnas.95.2.606

202 Behrens, S. J., & Parker, J. A. (2010). *Language in the Real World: An Introduction to Linguistics*. New York: Taylor & Francis.

203 Beker, H., & Piper, F. (1982). *Cipher Systems: The Protection of Communications*. London: Northwood Books.

204 Bernhardt, H. S. (2012). "The RNA World Hypothesis: The Worst Theory of the Early Evolution of Life (Except for All the Others)." *Biology Direct, 7*, 23.

205 Bettelheim, F. A., Brown, W. H., Campbell, M. K., Farrell, S. O., & Torres, O. J. (2010). *Introduction to Organic and Biochemistry* (8th ed.). Belmont, CA: Brooks/Cole.

206 Bird, R. J. (2003). *Chaos and Life: Complexity and Order in Evolution and Thought*. New York: Columbia University Press.

207 Cairns-Smith, A. G. (1982). *Genetic Takeover and the Mineral Origins of Life*. Cambridge, UK: Cambridge University Press.

208 Chiras, D. D. (2011). *Human Biology*. Burlington, MA: Jones & Bartlett Learning.

209 Conway Morris, S. (2000). "The Cambrian 'Explosion': Slow-Fuse or Megatonnage?" *Proceedings of the National Academy of Sciences of the United States of America, 97*, 4426–4429.

210 Cooper, G. M. (2000). "The Central Role of Enzymes as Biological Catalysts." In *The Cell: A Molecular Approach, American Society for Microbiology* (pp. 89–115). Sunderland, MA: Sinauer Associates. Retrieved from http://www.ncbi.nlm.nih.gov/books/NBK9921

211 Curtis, H., & Barnes, N. S. (1994). *Invitation to Biology*. New York: W. H. Freeman.

212 Emspak, J. (2011, October 14). "Scientists Build Self-Replicating Molecule." *Discovery .com*. Retrieved from http://news.discovery.com/tech/biotechnology/scientists-buil-self-replicating-molecule-111014.htm

213 Frank, A. (1997, November 1). "Quantum Honeybees." *Discover Magazine*. Retrieved from http://discovermagazine.com/1997/nov/quantumhoneybees1263

214 Frizzell, R. A. (1993). "The Molecular Physiology of Cystic Fibrosis." *Physiology, 8*, 117–120.

215 Fruton, J. S. (1999). *Proteins, Enzymes, Genes: The Interplay of Chemistry and Biology*. New Haven, CT: Yale University Press.

216 Gilbert, W. (1986). "Origin of Life: The RNA World." *Nature, 319*(6055), 618. doi:10.1038/319618a0

217 Gwynne, P., Begley, S., & Hager, M. (1979, August 20). "The Secrets of the Human Cell," *Newsweek, 94*, 48–54.

218 Hawking, S. (2007). *God Created the Integers: The Mathematical Breakthroughs That Changed History*. Philadelphia: Running Press.

219 Hazen, R. M. (2005). *Genesis: The Scientific Quest for Life's Origin*. Washington, DC: Joseph Henry Press.

220 Heijmans, B. T., Tobi, E. W., Stein, A. D., Putter, H., Blauw, G. J., Susser, E. S., . . . Lumey, L. H. (2008). "Persistent Epigenetic Differences Associated with Prenatal Exposure to Famine in Humans." *Proceedings of the National Academy of Sciences of the United States of America*, 105, 17046–17049.

221 Hirs, C. H., Timasheff, S. N., et al. (1964). *Progress in Nucleic Acid Research & Molecular Biology*. New York: Academic Press.

222 Horgan, J. (2011, February 28). "Pssst! Don't Tell the Creationists, but Scientists Don't Have a Clue How Life Began." *Cross-Check* [Blog]. Retrieved from http://blogs.scientificamerican.com/cross-check/2011/02/28/pssst-dont-tell-the-creationists-but-scientists-dont-have-a-clue-how-life-began/

223 Kauffman, S. (1996). *At Home in the Universe: The Search for the Laws of Self-Organization and Complexity*. New York: Oxford University Press.

224 Küppers, B. O. (1990). *Information and the Origin of Life*. Cambridge, MA: MIT Press.

225 "Life." (1974). In *Encyclopedia Britannica: Macropaedia*, pp. 893–894.

226 Dumont, M. L. (1873, April 3). "The Theory of Evolution in Germany." In N. Lockyer (Ed.), *Nature, 7*, 434.

227 Lowe, S. W., & Lin, A.W. (2000). "Apoptosis in Cancer." *Carcinogenesis, 21*, 485–495.

228 Lupo, A., Coyne, S., & Berendonk, T. U. (2012). "Origin and Evolution of Antibiotic Resistance: The Common Mechanisms of Emergence and Spread in Water Bodies." *Frontiers in Microbiology*, 3. Retrieved from http://www.ncbi.nlm.nih.gov/pmc/articles/PMC3266646

229 MacDonald, D. W. (2006). *The Encyclopedia of Mammals: Facts on File*. New York: Natural Science Library.

230 Maher, B. (2012, September 9). "Brendan Maher Writes about ENCODE/Junk DNA Publicity Fiasco." *Sandwalk* [Blog]. Retrieved from http://sandwalk.blogspot.com/2012/09/brendan-maher-writes-about-encodejunk.html

231 Margulis, L., & Chapman, M. J. (2009). *Kingdoms and Domains: An Illustrated Guide to the Phyla of Life on Earth*. New York: Academic Press.

232 Mendel, J. G. (1901). "Versuche über Pflanzenhybriden Verhandlungen des naturforschenden Vereines in Brünn, Bd. IV für das Jahr." *Journal of the Royal Horticultural Society, 26*, 1–32.

233 Meredith, S., Claybourne, A., Reiss, M. J., et al. (2006). *The Usborne Introduction to Genes & DNA: Internet Linked*. London: Usborne.

234 Merlo, L. M. F., Pepper, J. W., Reid, J., & Maley, C. C. (2006). "Cancer as an Evolutionary and Ecological Process." *Nature Reviews Cancer, 6*, 924–935.

235 Miller, S. L. (1953). "A Production of Amino Acids under Possible Primitive Earth Conditions." *Science, 117*, 528–529.

236 Mitchell, A., & Pilpel, Y. (2011). "A Mathematical Model for Adaptive Prediction of Environmental Changes by Microorganisms." *Proceedings of the National Academy of Sciences of the United States of America, 108*, 7271–7276.
237 Moore, R. C., Vodopich, D. S., & Cotner, S. H. (2011). *Arguing for Evolution: An Encyclopedia for Understanding Science*. Santa Barbara, CA: ABC-CLIO.
238 Nguyen, T., Brunson, D., Crespi, C. L., Penman, B. W., Wishnok, J. S., & Tannenbaum, S. R. (1992). "DNA Damage and Mutation in Human Cells Exposed to Nitric Oxide in Vitro." *Proceedings of the National Academy of Sciences of the United States of America, 89*(7), 3030–3034.
239 Nielsen, L. L., & Maneval, D. C. (1997). "P53 Tumor Suppressor Gene Therapy for Cancer." *Cancer Gene Therapy 5*(1), 52–63.
240 Pace, N. R. (2001). "The Universal Nature of Biochemistry." *Proceedings of the National Academy of Sciences of the United States of America, 98*, 805–808.
241 Patton, J. T. (2008). *Segmented Double-Stranded RNA Viruses: Structure and Molecular Biology*. Poole, UK: Horizon Scientific Press.
242 Reif, J. H. (2002). "Successes and Challenges." *Science, 296*(5567), 478–479.
243 Robinson, T. R. (2005). *Genetics for Dummies*. Indianapolis, IN: Wiley.
244 Riordan, J. R., Rommens, J. M., Kerem, B., Alon, N., Rozmahel, R., Grzelczak, Z., Zielinski, J., . . . Chou, J. L. (1989). "Identification of the Cystic Fibrosis Gene: Cloning and Characterization of Complementary DNA." *Science, 245*(4925), 1066–1073.
245 Rowland, M. (1992). *Biology*. Surrey, UK: Nelson.
246 Sanford, J. C. (2005). *Genetic Entropy & the Mystery of the Genome*. Waterloo, NY: FMS Publications.
247 Shannon, C.E.. "An Algebra for Theoretical Genetics." PhD dissertation, Cambridge, MA: Massachusetts Institute of Technology, 1940.
248 Shimkets, R. A. (2004). *Gene Expression Profiling: Methods and Protocols*. Totowa, NJ: Humana Press.
249 Smith, J., & Dixon, A. (2009, November 1). "Birds Do It. Bats Do It." *Greater Good*. Retrieved from http://greatergood.berkeley.edu/article/item/birds_do_it_bats_do_it
250 Szent-Gyoergyi, A. (1966). "Drive in Living Matter to Perfect Itself." *Journal of Individual Psychology, 22*(2), 153–162.
251 Ullmann, A. (2007). "Pasteur-Koch: Distinctive Ways of Thinking about Infectious Diseases." *Microbe, 2*(8), 383–387.
252 Von Neumann, J. (1951). "The General and Logical Theory of Automata." In L. A. Jeffress (Ed.), *Cerebral Mechanisms in Behavior* (pp. 1–41). Oxford, UK: Wiley.
253 Wark, K. (1995). *Advanced Thermodynamics for Engineers*. New York: McGraw-Hill.
254 Watson, J. D., & Berry, A. (2003). *DNA: The Secret of Life*. New York: Knopf.
255 Webber, M. A., & Piddock, L. J. V. (2003). "The Importance of Efflux Pumps in Bacterial Antibiotic Resistance." *Journal of Antimicrobial Chemotherapy, 51*, 9–11.
256 Wilkins, J. S. (2013). "The Salem Region: Two Mindsets about Science." In M. Pigliucci & M. Boudry (Eds.), *Philosophy of Pseudoscience: Reconsidering the Demarcation Problem* (pp. 397–416). Chicago: University of Chicago Press.

300 Genetic Code & Coding Theory

300 Bohlin, J., van Passel, M.W. J., Snipen, L., Kristoffersen, A. B., Ussery, D., & Hardy, S. P. (2012). "Relative Entropy Differences in Bacterial Chromosomes, Plasmids, Phages and Genomic Islands." *BMC Genomics, 13*, 66. doi:10.1186/1471-2164-13-66
301 Charrel, A. (1995). Tierra Network Version. *ATR Technical Report TR-H-145*.
302 Clayton, J., & Dennis, C. (Eds.). (2003). *50 Years of DNA*. Hampshire, UK: Palgrave Macmillan.
303 Crick, F. (1962, December 11). "Nobel Lecture: On the Genetic Code." Retrieved from http://www.nobelprize.org/nobel_prizes/medicine/laureates/1962/crick-lecture.html
304 Crick, F. (1968). "The Origin of the Genetic Code." *Journal of Molecular Biology, 38*, 367–379.

305 The ENCODE Project Consortium. (2011, April 19). "ENCODE: A User's Guide to the Encyclopedia of DNA Elements (ENCODE)." *PLoS Biology.* Retrieved from http://dx.plos.org/10.1371/journal.pbio.1001046

306 Freeland, S. J., & Hurst, L. D. (1998). "The Genetic Code Is One in a Million." *Journal of Molecular Evolution, 47*, 238–248.

307 Friedberg, E. C. W., & Siede, W. (1995). "DNA Repair and Mutagenesis." *Trends in Biochemical Sciences—Library Compendium, 20*, 440.

308 Garcia, J. A., & Jose, M. V. (2005). "Mathematical Properties of DNA Sequences from Coding and Noncoding Regions." *Revista Mexicana de Física, 51*, 122–130.

309 Gitschier, J. (2012). "It Was Heaven: An Interview with Evelyn Witkin." *PLoS Genetics* 8, e1003009.

310 Gitt, W. (1997). *In the Beginning Was Information*. Bielefeld, Germany: Christliche Literatur-Verbreitung.

311 Harold, F. M. (2001). *The Way of the Cell: Molecules, Organisms, and the Order of Life*. New York: Oxford University Press.

312 Kunkel, T. A., & Eric, D. A. (2005). "DNA Mismatch Repair." *Annual Review of Biochemistry, 74*, 681–710.

313 Lee, T. F. (1991). *The Human Genome Project: Cracking the Genetic Code of Life*. New York: Plenum.

314 Lipman, D. J., & Maizel, J. (1982). "Comparative Analysis of Nucleic Acid Sequences by Their General Constraints." *Nucleic Acids Research, 10*, 2723–2739.

315 Ohno, S. (1972). "So Much 'Junk' DNA in Our Genome." *Brookhaven Symposia in Biology, 23*, 366–370.

316 Perez, J.-C. (2010). "Codon Populations in Single-Stranded Whole Human Genome DNA Are Fractal and Fine-Tuned by the Golden Ratio 1.618." *Interdisciplinary Sciences*, 2(3), 228–240.

317 Pressing, J., & Reanney, D. C. (1984). "Divided Genomes and Intrinsic Noise." *Journal of Molecular Evolution, 20*, 135–146.

318 Ricard, J. (2006). *Emergent Collective Properties, Networks, and Information in Biology*. New York: Elsevier.

319 Sadovsky, M. G. (2006). "Information Capacity of Nucleotide Sequences and Its Applications." *Bulletin of Mathematical Biology, 68*, 785–806.

320 Shannon, C. E. (1948). "A Mathematical Theory of Communication." *Bell System Technical Journal, 27*, 379–423.

321 Sklar, B. (2001). *Digital Communications* (2nd ed.). New York: Prentice Hall.

322 Walton, M. (2012, December 4). "Mice, Men Share 99 Percent of Genes." *CNN.com.* Retrieved from http://articles.cnn.com/2002-12-04/tech/coolsc.coolsc.mousegenome_1_human-genome-new-human-genes-genes-that-cause-disease?_s=PM:TECH

323 Wiener, N. (1961). *Cybernetics: Or Control and Communication in the Animal and the Machine*, 25. Cambridge, MA: MIT Press.

324 Wiener, N. (1965). *Cybernetics: Or Control and Communication in the Animal and the Machine*, 25. Cambridge, MA: MIT Press.

325 Yockey, H. P. (2000). "Origin of Life on Earth and Shannon's Theory of Communication." *Computers & Chemistry, 24*, 105–123.

326 Yockey, H. P. (2005). *Information Theory, Evolution, and the Origin of Life*. Cambridge, UK: Cambridge University Press.

400 Linguistic Models of DNA

400 Bassler, B. L. (2002). "Small Talk: Cell-to-Cell Communication in Bacteria." *Cell, 109*, 421–424.

401 Ben Jacob, E., Becker, I., Shapira, Y., & Levine, H. (2004). "Bacterial Linguistic Communication and Social Intelligence." *Trends in Microbiology, 12*(8), 366–372.

402 Jerne, N. K. (1985). "The Generative Grammar of the Immune System." *Bioscience Reports, 5*, 439–451.
403 Ji, S. (1999). "The Linguistics of DNA: Words, Sentences, Grammar, Phonetics, and Semantics." *Annals of the New York Academy of Sciences, 870*, 411–417.
404 Katz, G. (2008). "The Hypothesis of a Genetic Protolanguage: An Epistemological Investigation." *Biosemiotics*, 1, 57–73.
405 Searls, D. B. (2002). "The Language of Genes." *Nature, 420*, 211–217.
406 Smith, A. D. (1997). *Oxford Dictionary of Biochemistry and Molecular Biology*. New York: Oxford University Press.
407 Witzany, G. (2008). "Bio-Communication of Bacteria and Their Evolutionary Roots in Natural Genome Editing Competences of Viruses." *Open Evolution Journal*, 2, 44–54.

500 Cellular Cognition

500 Andrianantoandro, E., Basu, S., Karig, D. K., & Weiss, R. (2006, May 16). "Synthetic Biology: New Engineering Rules for an Emerging Discipline." *Molecular Systems Biology*, 2:2006.0028.
501 Baluška, F., Mancuso, S., Volkmann, D., & Barlow, P. W. (2009). "The 'Root-Brain' Hypothesis of Charles and Francis Darwin: Revival after More Than 125 Years." *Plant Signaling & Behavior, 4*, 1121–1127.
502 Chowdhury, D., Nishinari, K., & Schadschneider, A. (2004). "Self-Organized Patterns and Traffic Flow in Colonies of Organisms: From Bacteria and Social Insects to Vertebrates." *Phase Transitions, 77*, 601–624.
503 Duke, R. C., Ojcius, D. M., & Young, J. D. E. (1996). "Cell Suicide in Health and Disease." *Scientific American, 275*, 80–87.
504 Dworkin, M. (1983). "Tactic Behavior of Myxococcus xanthus." *Journal of Bacteriology, 154*, 452–459.
505 Engelberg-Kulka, H., Amitai, S., Kolodkin, I., & Hazan, R. (2006). "Bacterial Programmed Cell Death and Multicellular Behavior in Bacteria." *PLoS Genetics*, 2, e135.
506 Fraser, C. M., Gocayne, J. D., White, O., Adams., M. D., Clayton, R. A., Fleischmann, R. D., Bult, C. J., . . . Venter, J. C. (1995). "The Minimal Gene Complement of Mycoplasma genitalium." *Science, 270*(5235), 397–403.
507 Golden, J. W., Robinson, S. J., & Haselkorn, R. (1985). "Rearrangement of Nitrogen Fixation Genes during Heterocyst Differentiation in the Cyanobacterium Anabaena." *Nature, 314*, 419–423.
508 Hoffmeyer, J. (2000). "The Biology of Signification." *Perspectives in Biology and Medicine, 43*, 252–268.
509 Lindqvist, A., Rodriguez, V., & Medema, R. H. (2009). "The Decision to Enter Mitosis: Feedback and Redundancy in the Mitotic Entry Network." *The Journal of Cell Biology, 185*, 193–202.
510 Liu, R., & Ochman, H. (2007). "Origins of Flagellar Gene Operons and Secondary Flagellar Systems." *Journal of Bacteriology, 189*, 7098–7104.
511 Li, Z., Rosenbaum, M. A., Venkataraman, A., Tam, T. K., Katz, E., & Angenent, L. T. (2011). "Bacteria-Based AND Logic Gate: A Decision-Making and Self-Powered Biosensor." *Chemical Communications, 47*, 3060–3062.
512 Loewenstein, W. R. (1999). *The Touchstone of Life: Molecular Information, Cell Communication, and the Foundations of Life*. New York: Oxford University Press.
513 Lyon, P. (2007). "From Quorum to Cooperation: Lessons from Bacterial Sociality for Evolutionary Theory." *Studies in History and Philosophy of Biological and Biomedical Sciences, 38*, 820–833.
514 Martincorena, I., Seshasayee, A. S. N., & Luscombe, N. M. (2012). "Evidence of Non-Random Mutation Rates Suggests an Evolutionary Risk Management Strategy." *Nature, 485*, 95–98.
515 McMenamin, G. R. (2002). *Forensic Linguistics: Advances in Forensic Stylistics*. Boca Raton, FL: CRC.

516 Miller, M. B., & Bassler, B. L. (2001). "Quorum Sensing in Bacteria." *Annual Reviews in Microbiology, 55*, 165–199.
517 Morita, Y., Kataoka, A., Shiota, S., Mizushima, T., & Tsuchiya, T. (2000). "NorM of *Vibrio parahaemolyticus* Is an Na+-Driven Multidrug Efflux Pump." *Journal of Bacteriology, 182*, 6694–6697.
518 Nakagaki, T. (2001). "Smart Behavior of True Slime Mold in a Labyrinth." *Research in Microbiology, 152*, 767–770.
519 Nakagaki, T., Kobayashi, R., Nishiura, Y., & Ueda, T. (2004). "Obtaining Multiple Separate Food Sources: Behavioural Intelligence in the Physarum Plasmodium." *Proceedings of the Royal Society of London. Series B: Biological Sciences, 271*, 2305–2310.
520 Neuman, Y. (2008). *Reviving the Living: Meaning Making in Living Systems*. New York: Elsevier.
521 Perkins, T. J., & Swain, P. S. (2009). "Strategies for Cellular Decision-Making." *Molecular Systems Biology, 5*(1), 326.
522 Shapiro, J. A. (1988). "Bacteria as Multicellular Organisms." *Scientific American, 258*, 82–89.
523 Shapiro, J. A., & Dworkin, M. (1997). *Bacteria as Multicellular Organisms*. New York: Oxford University Press.
524 Sudo, S. Z., & Dworkin, M. (1973). "Comparative Biology of Prokaryotic Resting Cells." *Advances in Microbial Physiology, 9*, 153–224.
525 Zohar, A., & Ginossar, S. (1998). "Lifting the Taboo Regarding Teleology and Anthropomorphism in Biology Education—Heretical Suggestions." *Science Education*, 82, 679–697.

600 Post-Darwinian Models of Evolution

600 Adhya, S. L., & Shapiro, J. A. (1969). "The Galactose Operon of E. COLI K-12. I. Structural and Pleiotropic Mutations of the Operon." *Genetics, 62*, 231–247.
601 Ahmadjian, V. (1962). "Investigations on Lichen Synthesis." *American Journal of Botany*, 49(3), 277–283.
602 Alkan, C., Ventura, M., Archdiacono, N., Rocchi, M., Sahinalp, S. C., & Eichler, E. E. (2007). "Organization and Evolution of Primate Centromeric DNA from Whole-Genome Shotgun Sequence Data." *PLoS Computational Biology, 3*, 1807–1818.
603 Barbash, D. A. (2010). "Ninety Years of *Drosophila melanogaster* Hybrids." *Genetics, 186*, 1–8.
604 Bassler, B. L. (2009, February). "How Bacteria 'Talk.'" *TED2009*. Retrieved from http://www.ted.com/talks/bonnie_bassler_on_how_bacteria_communicate.html
605 Randal Bollinger, R., Barbas, A. S., Bush, E. L., Lin, S. S., & Parker, W. (2007). "Biofilms in the Large Bowel Suggest an Apparent Function of the Human Vermiform Appendix." *Journal of Theoretical Biology, 249*, 826–831.
606 Bradt, S. (2009, October 8). "A Look Inside: Scientists Have Deciphered 3-D Structure of the Human Genome." *Harvard Gazette*. Retrieved from http://news.harvard.edu/gazette/story/2009/10/3-d-human-genome/
607 Brandon, R. A. (1977). "Interspecific Hybridization among Mexican and United States Salamanders of the Genus Ambystoma Under Laboratory Conditions." *Herpetologica*, 33(2), 133–152.
608 Bray, D. (2009). *Wetware: A Computer in Every Living Cell*. New Haven, CT: Yale University Press.
609 Cavalier-Smith, T. (2003). "Microbial Muddles." *BioScience, 53*, 1008–1013.
610 Chamovitz, D. (2012). *What a Plant Knows: A Field Guide to the Senses*. New York: Macmillan.
611 Conova, S. (2014, February 5). "Drawing Back DNA 'Curtains' on New Gene-Editing Method." *Columbia University Medical Center Newsroom*. Retrieved from http://newsroom.cumc.columbia.edu/blog/2014/02/05/Dna-curtains-draw-back-secrets-new-gene-editing-method/
612 Diekmann, Y., & Pereira-Leal, J. B. "Evolution of Intracellular Compartmentalization," *Biochemical Journal, 449*(2), 319–331.

613 Duharcourt, S., Lepère, G., & Meyer, E. (2009). "Developmental Genome Rearrangements in Ciliates: A Natural Genomic Subtraction Mediated by Non-Coding Transcripts." *Trends in Genetics, 25*(8), 344–350. Retrieved from http://www.ncbi.nlm.nih.gov/pubmed/19596481

614 Eldredge, N., & Gould, S. J. (1972). "Punctuated Equilibria: An Alternative to Phyletic Gradualism." In T. J. M. Schopf (Ed.), *Models in Paleobiology* (pp. 82–115). San Francisco: Freeman, Cooper.

615 Fall, S., Mercier, A., Bertolla, F., Calteau, A., Gueguen, L., Perrière, G., . . . Simonet, P. (2007). "Horizontal Gene Transfer Regulation in Bacteria as a 'Spandrel' of DNA Repair Mechanisms." *PLoS One* 2: e1055. Retrieved from http://journals.plos.org/plosone/article?id=10.1371/journal.pone.0001055

616 Gould, S. J. (1977). "Evolution's Erratic Pace." *Natural History, 86*, 12–16.

617 Gould, S. J. (1980). "Is a New and General Theory of Evolution Emerging?" *Paleobiology*, *6*(1), 119–130.

618 Gould, S. J. (2002). *Rocks of Ages*. New York: Vintage/Ebury.

619 Grant, T., & Woods, A. (2002). *Reason in Revolt: Dialectical Philosophy and Modern Science*. New York: Algora.

620 Gray, T. A. (2013). "Distributive Conjugal Transfer in *Mycobacteria* Generates Progeny with Meiotic-Like Genome-Wide Mosaicism, Allowing Mapping of a Mating Identity Locus," *PLoS Biology,* 11: e1001602.

621 Ho, M. W. (n.d.). "Horizontal Gene Transfer." *Institute of Science in Society.* Retrieved from http://online.sfsu.edu/rone/GEessays/horizgenetransfer.html

622 Holt, S., & Aguirre, L. (2007, July 24). "Epigenetics" [Video and Transcript]. *Nova.* Retrieved from http://www.pbs.org/wgbh/nova/body/epigenetics.html

623 Horgan, J. (1997). *The End of Science: Facing the Limits of Knowledge in the Twilight of the Scientific Age*. Reading, MA: Addison-Wesley.

624 Jeon, K. W. (1987). "Change of Cellular 'Pathogen' into Required Cell Components." *Annals of the New York Academy of Sciences, 503*, 359–371.

625 Jeon, T. J., & Jeon, K. W. (2004). "Gene Switching in Amoeba Proteus Caused by Endosymbiotic Bacteria." *Journal of Cell Science, 117*(4), 535–543.

626 Kasahara, M. (2007). "The 2R Hypothesis: An Update." *Current Opinion in Immunology, 19*, 547–552.

627 Keller, E. F. (1984). *A Feeling for the Organism, 10th Aniversary Edition: The Life and Work of Barbara McClintock*. New York: Times Books.

628 Kelly, M. "Far from Random, Evolution Follows a Predictable Genetic Pattern, Princeton Researchers Find." Princeton University News Archive. Retrieved from http://www.princeton.edu/main/news/archive/S35/06/74S40/

629 Kolata, G. "Bits of Mystery Data, Far from 'Junk,' Play Crucial Role." *The New York Times.* Retrieved from http://www.nytimes.com/2012/09/06/science/far-from-junk-dna-dark-matter-proves-crucial-to-health.html

630 Lindblad-Toh, K., Garber, M., Zuk, O., Lin, M. F., Parker, B. J., Washietl, S., Kheradpour, P., . . . Kellis, M. (2011). "A High-Resolution Map of Human Evolutionary Constraint Using 29 Mammals." *Nature, 478*, 476–482.

631 Liu, R., & Ochman, H. (2007). "Stepwise Formation of the Bacterial Flagellar System." *Proceedings of the National Academy of Sciences of the United States of America, 104*, 7116–7121.

632 Lokki, J., Suomalainen, E., Saura, A., & Lankinen, P. (1975). "Genetic Polymorphism and Evolution in Parthenogenetic Animals. II. Diploid and Polyploid *SOLENOBIA TRIQUETRELLA* (Lepidoptera: Psychidae)." *Genetics, 79*, 513–525.

633 Lutz, B., Lu, H. C., Eichele, G., Miller, D., & Kaufman, T. C. (1996). "Rescue of Drosophila Labial Null Mutant by the Chicken Ortholog Hoxb-1 Demonstrates That the Function of Hox Genes Is Phylogenetically Conserved." *Genes & Development, 10*(2), 176–184. doi:10.1101/gad.10.2.176.

634 Lyne, J., & Howe, H. F. (1986). "'Punctuated Equilibria': Rhetorical Dynamics of a Scientific Controversy." *Quarterly Journal of Speech,* 72, 132–147.

635 Mann, C. (1991). "Lynn Margulis: Science's Unruly Earth Mother." *Science, 252*, 378–381.
636 Margulis, L., & Fester, R. (1991). *Symbiosis as a Source of Evolutionary Innovation: Speciation and Morphogenesis*. Cambridge, MA: MIT Press.
637 Margulis, L., & Sagan, D. (2003). *Acquiring Genomes: A Theory of the Origin of Species*. New York: Basic Books.
638 Lynn Margulis is quoted in chapter 5 of Horgan, J. (1997). *The End of Science: Facing the Limits of Knowledge in the Twilight of the Scientific Age* (pp. 140–141). Reading, MA: Addison-Wesley.
639 McClintock, B. (1950). "The Origin and Behavior of Mutable Loci in Maize." *Proceedings of the National Academy of Sciences of the United States of America, 36*, 344–355.
640 McClintock, B. (1983, December 8). "Nobel Lecture: The Significance of Responses of the Genome to Challenge." Retrieved from http://www.nobelprize.org/nobel_prizes/medicine/laureates/1983/mcclintock-lecture.html
641 Meyer, A., & dePeer, Y. (2005). "From 2R to 3R: Evidence for a Fish-Specific Genome Duplication (FSGD)." *BioEssays, 27*, 937–945.
642 Comfort, N. C. (2003). *The Tangled Field: Barbara McClintock's Search for the Patterns of Genetic Control*. Cambridge, MA: Harvard University Press.
643 Noble, D. (2015). "Evolution Beyond Neo-Darwinism: A New Conceptual Framework." *The Journal of Experimental Biology, 218*(1), 7–13.
644 Noble, D. (2008). *The Music of Life: Biology beyond Genes*. New York: Oxford University Press.
645 Noble, D. (2013). "Physiology Is Rocking the Foundations of Evolutionary Biology." *Experimental Physiology, 98*(8), 1235–1243.
646 Noble, D. "Physiology and the Revolution in Evolutionary Biology." Keynote speech at the International Conference of Physiological Sciences in 2012, Suzhou, China. *VOX: Voices from Oxford* [Video]. Retrieved from http://www.voicesfromoxford.org/video/physiology-and-the-revolution-in-evolutionary-biology/184
647 Novick, R. P. (1980). "Plasmids." *Scientific American, 243*, 102.
648 Ohno, S., et al. (1970). *Evolution by Gene Duplication*. London: George Allen & Unwin.
649 Pennisi, E. (2012, September 7). "ENCODE Project Writes Eulogy for Junk DNA." *Science, 337*, 1159–1161.
650 Prescott, D. M. (2000). "Genome Gymnastics: Unique Modes of DNA Evolution and Processing in Ciliates." *Nature Reviews Genetics, 1*, 191–198.
651 Rapp, R. A., & Wendel, J. F. (2005). "Epigenetics and Plant Evolution." *New Phytologist, 168*, 81–91.
652 Ryan, F. (2011). *Virolution*. New York: HarperCollins.
653 Ryan, F. (2002). *Darwin's Blind Spot: Evolution beyond Natural Selection*. New York: Houghton Mifflin Harcourt.
654 Sapp, J., Carrapico, F., & Zolotonosov, M. (2002). "Symbiogenesis: The Hidden Face of Constantin Merezhkowsky." *History and Philosophy of the Life Sciences, 24*, 413–440.
655 Schmidt, M. (2009). *Beyond Antibiotics: Strategies for Living in a World of Emerging Infections and Antibiotic-Resistant Bacteria*. Berkeley, CA: North Atlantic Books.
656 Schrödinger, Erwin. (1944). *What Is Life—The Physical Aspect of the Living Cell*. Cambridge, UK: Cambridge University Press.
657 Shapiro, J. A. (2005). "Barbara McClintock, 1902–1992." *BioEssays, 14*, 791–792.
658 Shapiro, J. A. (2009). "Letting Escherichia Coli Teach Me about Genome Engineering." *Genetics, 183*, 1205–1214.
659 Shapiro, J. A. (2011). *Evolution: A View from the 21st Century*. Upper Saddle River, NJ: FT Press.
660 Shapiro, J. A. (2012, March 25). "DNA as Poetry: Multiple Messages in a Single Sequence." *The Huffington Post*. Retrieved from http://www.huffingtonpost.com/james-a-shapiro/dna-as-poetry-multiple-me_b_1229190.html
661 Shapiro, J. A. (2012, April 7). "Purposeful, Targeted Genetic Engineering in Immune System Evolution." *The Huffington Post*. Retrieved in 2012 from http://

www.huffingtonpost.com/james-a-shapiro/genetic-engineering-immune-system-evolution_b_1255771.html

662 Shapiro, J. A. (2012, May 19). "Cell Cognition and Cell Decision Making." *The Huffington Post*. Retrieved from http://www.huffingtonpost.com/james-a-shapiro/cell-cognition_b_1354889.html

663 Shapiro, J. A. (1999). "Genome System Architecture and Natural Genetic Engineering in Evolution." *Annals of the New York Academy of Sciences, 870*(1), 23–35.

664 Shapiro, J. A. (2013). "How Life Changes Itself: The Read–Write (RW) Genome." *Physics of Life Reviews, 10*(3), 287–323.

665 Shapiro, J. A. (2009). "Revisiting the Central Dogma in the 21st Century." *Annals of the New York Academy of Sciences, 1178*, 6–28. doi:10.1111/j.1749-6632.2009.04990.x

666 Shapiro, J. A. (2013, January 30). "Why the 'Gene' Concept Holds Back Evolutionary Thinking." *The Huffington Post*. Retrieved from http://www.huffingtonpost.com/james-a-shapiro/why-the-gene-concept-hold_b_2207245.html

667 Singer, E. (2009, February 4). "A Comeback for Lamarckian Evolution?" *Technology Review*. Retrieved from http://www.technologyreview.com/news/411880/a-comeback-for-lamarckian-evolution

668 Smith, D. C., & Douglas, A. E. (1987). *The Biology of Symbiosis*. London: Edward Arnold.

669 Somit, A., & Peterson, S. A. (1992). *The Dynamics of Evolution: The Punctuated Equilibrium Debate in the Natural and Social Sciences*. Ithaca, NY: Cornell University Press.

670 "Spider DNA Spurs Search into Arachnid Secrets." (2014, May 6). *PhysOrg*. Retrieved from http://phys.org/news/2014-05-spider-dna-spurs-arachnid-secrets.html

671 Stadler, D., & Moyer, R. (1981). "Induced Repair of Genetic Damage in Neurospora." *Genetics, 98*, 763–774.

672 Starrett, J., Garb, J. E., Kuelbs, A., Azubuike, U. O., & Hayashi, C. Y. (2012) "Early Events in the Evolution of Spider Silk Genes," *PLoS One 7*(6), e38084. Retrieved from http://journals.plos.org/plosone/article?id=10.1371/journal.pone.0038084

673 Stebbins, G. L. (1951). "Cataclysmic Evolution." *Scientific American, 184*, 54–59.

674 Syvanen, M., & Kado, C. I. (2002). *Horizontal Gene Transfer*. New York: Academic Press.

675 Talbott, S. L. (2012, July 3). "Biology Worthy of Life: Getting Over the Code Delusion." Retrieved from http://natureinstitute.org/txt/st/mqual/genome_4.htm

676 Taylor, G. R. (1983). *The Great Evolution Mystery*. HarperCollins.

677 Thomas, E. A. (1939). *Uber die Biologie von Flechtenbildnern* (9th ed.). Beiträge zur Kryptogamenflora der Schweiz.

678 Trafton, A. (2009, October 9) "A New Dimension for Genome Studies." *MIT News*. Retrieved from http://web.mit.edu/newsoffice/2009/3d-genome.html

679 ScienceDaily. (2012, July 3). "Two Species Fused to Give Rise to Plant Pest a Few Hundred Years Ago." Retrieved from http://www.sciencedaily.com/releases/2012/07/120703133725.htm

680 Woese, C. R. (2004). "A New Biology for a New Century." *Microbiology and Molecular Biology Reviews, 68*, 173–186.

681 Woese, C. R., & Fox, G. E. (1977). "Phylogenetic Structure of the Prokaryotic Domain: The Primary Kingdoms." *Proceedings of the National Academy of Sciences of the United States of America, 74*, 5088–5090.

682 Yandell, K. (2013, July 9). "Bacterial Gene Transfer Gets Sexier." *The Scientist*. Retrieved from http://www.the-scientist.com/?articles.view/articleNo/36410/title/Bacterial-Gene-Transfer-Gets-Sexier/

683 Zhang, T. Y., & Meaney, M. J. (2010). "Epigenetics and the Environmental Regulation of the Genome and Its Function." *Annual Review of Psychology, 61*, 439–466.

700 Technology

700 Adami, C., & Brown, C. T. (1994). "Evolutionary Learning in the 2D Artificial Life System 'Avida.'" In R. A. Brooks & P. Maes (Eds.), *Artificial Life IV: Proceedings of the Fourth International Workshop on the Synthesis and Simulation of Living Systems* (pp. 377–381). Cambridge, MA: MIT Press.

701 Altshuller, G. S., Shulyak, L., & Rodman, S. (1997). *40 Principles: TRIZ Keys to Innovation*. Worcester, MA: Technical Innovation Center.

702 Banzal, S. (2007). *Data and Computer Network Communication*. Bangalore: Firewall Media/Laxmi Publications.

703 FoxBlitzz. (2006, August 11). "Corrupting Super Mario Bros" [Video]. Retrieved from http://www.youtube.com/watch?v=gN0sY0rX8rI&feature=youtube

704 Holland, J. H. (1975). *Adaptation in Natural and Artificial Systems: An Introductory Analysis with Applications to Biology, Control, and Artificial Intelligence*. Ann Arbor, MI: University of Michigan Press.

705 Jaki, S. L. (1969). *Brain, Mind, and Computers*. New York: Herder & Herder.

706 Koza, J. R., Bennett, F. H., Andre, D., & Keane, M. A. (2003). "Genetic Programming: Biologically Inspired Computation That Creatively Solves Non-Trivial Problems." In L. F. Landweber & E. Winfree (Eds.), *Evolution as Computation: DIMACS Workshop* (pp. 95–124). Berlin: Springer-Verlag.

707 Kurzweil, R. (2005). *The Singularity Is Near: When Humans Transcend Biology*. New York: Viking Adult.

708 Marshall, P. S., & Rinaldi, J. S. (2005). *Industrial Ethernet: How to Plan, Install, and Maintain TCP/IP Ethernet Networks: The Basic Reference Guide for Automation and Process Control Engineers*. Research Triangle Park, NC: ISA.

709 Ray, T. S. (1991). *Scientific Excellence in Supercomputing: The IBM 1990 Contest Prize Papers*. Athens, GA: Baldwin Press.

710 Ray, T. S. (1992). *Evolution, Ecology and Optimization of Digital Organisms* (Report No. 92-08-942). Santa Fe, NM: Santa Fe Institute.

711 Runyon, J. (2012, August 2). "An Unexpected Ass Kicking" [Blog post]. Retrieved from http://joelrunyon.com/two3/an-unexpected-ass-kicking

712 Stranneby, D., & Walker, W. (2004). *Digital Signal Processing and Applications*. Oxford, UK: Newness.

713 Turing, A. M. (1950). "Computing Machinery and Intelligence." *Mind, 59*, 433–460.

714 Watson, R. A., & Pollack, J. B. (2002). "A Computational Model of Symbiotic Composition in Evolutionary Transitions." *Biosystems, 69*, 187–209.

800 Mathematics

800 Chaitin, G. J. (1990). *Information Randomness & Incompleteness: Papers on Algorithmic Information Theory*. Teaneck, NJ: World Scientific Publishing Company Incorporated. Reprinted from Chaitin, G. J. (1975). "Randomness and Mathematical Proof." *Scientific American*, 232(5), 47–52.

801 Gabbay, D. M., Thagard, P., Woods, J., Bandyopadhyay, P. S., & Forster, M. R. (2011). *Philosophy of Statistics*, 7. New York: Elsevier.

802 Pellionisz, A. J., Graham, R., Pellionisz, P. A., & Perez, J. (2013). "Recursive Genome Function of the Cerebellum: Geometric Unification of Neuroscience and Genomics." In *Handbook of the Cerebellum and Cerebellar Disorders*, 1381–1423. Springer Netherlands.

900 History, Philosophy, & Theology

900 "Science and Religion" [Special issue]. (1999, July–August). *The Skeptical Inquirer*, 23(4).

901 *Holy Bible: New International Translation*. (1984). Grand Rapids, MI: Zondervan.

902 Boyle, R. (1989). "Of the Excellency and Grounds of the Corpuscular or Mechanical Philosophy." In M. R. Matthews, *The Scientific Background to Modern Philosophy: Selected Readings* (pp. 109–123). Indianapolis: Hackett. (Original work published 1674.)
903 Brown, N. M. (2011, January 3). "Everything You Think You Know about the Dark Ages Is Wrong" [Blog post]. Retrieved from http://religiondispatches.org/everything-you-think-you-know-about-the-dark-ages-is-wrong/
904 Butler, S. (1879). *Evolution, Old and New*. London: Harwick & Bogue.
905 Clark, R. W. (1984). *Einstein: The Life and Times*. New York: Avon.
906 Collins, F. S. (2006). *The Language of God: A Scientist Presents Evidence for Belief*. New York: Simon & Schuster.
907 Coperincus, N. (1543). "Dedication (Excerpts)." In *On the Revolutions of the Heavenly Bodies*. Retrieved from http://www.historyguide.org/earlymod/dedication.html
908 Darling, D. (2004). *The Universal Book of Mathematics: From Abracadabra to Zeno's Paradoxes*. Hoboken, NJ: Wiley.
909 Darlington, C. D. (1977). "Obituary: TD Lysenko." *Nature, 266*, 287–288.
910 Davies, P. (1993). *Mind of God: The Scientific Basis for a Rational World*. New York: Simon & Schuster.
911 Davies, P. "Stephen Hawking's Big Bang Gaps." (2010, September 4). *The Guardian UK*. Retrieved from http://www.guardian.co.uk/commentisfree/belief/2010/sep/04/stephen-hawking-big-bang-gap
912 Dilley, S. (2012). "Charles Darwin's Use of Theology in the Origin of Species." *British Journal for the History of Science, 45*, 29–56.
913 Einstein, A. (1934). *Mein Weltbild*. Amsterdam: Querido Verlag.
914 Oxford Dictionaries. (2015). "The OEC: Facts about the Language." Retrieved from http://oxforddictionaries.com/words/the-oec-facts-about-the-language
915 Ferngren, G. B. (2002). *Science and Religion: A Historical Introduction*. Baltimore, MD: Johns Hopkins University Press.
916 Gaither, C., & Cavazos-Gaither, A. E. (2012). *Gaither's Dictionary of Scientific Quotations: A Collection of Approximately 27,000 Quotations Pertaining to Archaeology, Architecture, Astronomy, Biology, Botany, Chemistry, Cosmology, Darwinism, Engineering, Geology, Mathematics, Medicine, Nature, Nursing, Paleontology, Philosophy, Physics, Probability, Science, Statistics, Technology, Theory, Universe, and Zoology*. New York: Springer.
917 Galileo, G. (1957). "Letter to Madame Christina of Lorraine, Grand Duchess of Tuscany, Concerning the Use of Biblical Quotations in Matters of Science." In S. Drake (Ed. and Trans.), *Discoveries and Opinions of Galileo* (pp. 175–216). New York: Doubleday.
918 Greer, R., Trans. (1979). *Origen: An Exhortation to Martyrdom, Prayer and Selected Works. Classics of Western Spirituality*. Mahwah, NJ: Paulist Press.
919 Gies, J. (1995). *Cathedral, Forge and Waterwheel*. New York: Harper Perennial.
920 Gigot, F. (1912). "Book of Wisdom." In *The Catholic Encyclopedia*. New York: Robert Appleton. Retrieved from http://www.newadvent.org/cathen/15666a.htm
921 Godart, O., & Heller, M. (1985). "Cosmology of Lemaître." Tucson, AZ: Pachart.
922 Grcic, J., & Grcic, P. D. J. (2009). *Facing Reality: An Introduction to Philosophy, Revised Edition*. Bloomington, IN: AuthorHouse.
923 Ham, K., & Hodge, B. (2011). *How Do We Know the Bible Is True?* Green Forest, AR: Master Books.
924 Hawking, S., & Mlodinow, L. (2010). *The Grand Design*. London: Transworld Digital.
925 Hitchens, Peter. *The Rage Against God*. Bloomsbury, 2010.
926 Hume, D. (1890). *A Treatise of Human Nature: Being an Attempt to Introduce the Experimental Method of Reasoning into Moral Subjects; And, Dialogues Concerning Natural Religion*. London: Longmans, Green.
927 Jaki, S. L. (1978). *The Road of Science and the Ways to God*. Chicago: University of Chicago Press.
928 Jammer, M. (2002). *Einstein and Religion: Physics and Theology*. Princeton, NJ: Princeton University Press.
929 Johnson, P. (2012). *History of Christianity*. New York: Simon & Schuster.

930 Kaplan, A., & Elman, Y. (1993). *Immortality, Resurrection, and the Age of the Universe: A Kabbalistic View*. Jersey City, NJ: Ktav.
931 Kepler, J. (1596). *Mysterium Cosmographicum*.
932 Kramer, M. (Ed.). (1999). *The Black Book of Communism: Crimes, Terror, Repression*. Cambridge, MA: Harvard University Press.
933 Moreland, J. P., & Craig, W. L. (2003). *Philosophical Foundations for a Christian Worldview*. Downer's Grove, IL: IVP Academic.
934 Nagel, T. (2012). *Mind and Cosmos: Why the Materialist Neo-Darwinian Conception of Nature Is Almost Certainly False*. New York: Oxford University Press.
935 Newton, I. (1714). *Philosophiæ Naturalis Principia Mathematica*. Sumptibus Societatis.
936 Newton, I. *Extract from Untitled Treatise on Revelation (section 1.1)*. Yahuda Ms edition. National Library of Israel, Jerusalem, Israel, f. 14r.9.
937 Philipse, H. (2012). *God in the Age of Science? A Critique of Religious Reason*. New York: Oxford University Press.
938 Pigliucci, M. (2012). "Who Knows What: The War between Science and the Humanities." Aeon. Retrieved from http://philpapers.org/rec/PIGWKW
939 Planck, M. (1955). *Religion und naturwissenschaft, vortrag gehalten im mai 1937*. Leipzig, Germany: JA Barth.
940 Piatt, C. (2014, July 11). "Frank Schaeffer: The God-Believing Atheist." *The Huffington Post*. Retrieved from http://www.huffingtonpost.com/christian-piatt/frank-schaeffer-the-god-b_b_5306149.html
941 Potts, D. (1996). *Mesopotamian Civilization: The Material Foundations*. Ithaca, NY: Cornell University Press.
942 Russell, B. (1919). *Mysticism and Logic: And Other Essays*. New York, London: Longmans, Green.
943 Stark, R. (2005). *The Victory of Reason: How Christianity Led to Freedom, Capitalism, and Western Success*. New York: Random House.
944 Stark, R. (2011). *The Triumph of Christianity: How the Jesus Movement Became the World's Largest Religion*. New York: HarperOne.
945 Ussher, J., Pierce, L., & Pierce, M. (Eds.) (2003). *Annals of the World* (Vol. 1). Green Forest, AR: New Leaf.
946 Walton, J. H. (2010). *The Lost World of Genesis One: Ancient Cosmology and the Origins Debate*. Downer's Grove, IL: IVP Academic.
947 Wells, P. S. (2009). *Barbarians to Angels: The Dark Ages Reconsidered*. New York: W. W. Norton.
948 O'Brian, N. F. (2006). "What's a 13th-Century Pope Got to Do with Stem Cells? Nothing at All." *Catholic News Service*. Retrieved from http://www.catholicnews.com/data/stories/cns/0604116.htm
949 Whitefield, R. (2006, June 12). "The Hebrew Word 'Yom' Used with a Number in Genesis 1. What Does 'Yom' Mean in Genesis 1?" Retrieved from http://www.godandscience.org/youngearth/yom_with_number.pdf
950 Winsley, J. (2005). *Why Do They That Know Him Not See His Days? Studies in God's Millennial Week*. Hokitika, New Zealand: Touching the King Publications.
951 Dimitrov, T. (2010, March). "Part I. 50 Nobel Laureates Who Believe in GOD: Nobel Scientists (1)." *Scientific GOD Journal, 1*(3), 151, 263.

Illustrations on pages 42, 44, 53, 67, 69, 95, 99, 100, 138, 139, 185, 201, 244, 297, and 342 by Kristin Mount
Illustrations on pages 87, 98, 136, and 146 by Danielle Flanagan
Illustrations on pages 123, 124, 126, 127, and 128 by Leslie McGrath
Photo on page 14 © Tim Evanson (www.flickr.com/photos/23165290@N00/6671114439/)
Photo on page 31 © Thomas Kaufman, PhD, University of Indiana
Photo on page 61 © Joel Runyon, used by permission
Cartoon on page 94 © Nick D. Kim, scienceandink.com, used by permission
Image on page 104 (top) from the RCSB Protein Data Bank (www.rcsb.org) of PDB ID 1QTJ (A. K. Shrive, A. M. Metcalfe, J. R. Cartwright, & T. J. Greenhough (1999), "Crystal Structure of Limulus Polyphemus SAP," *Journal of Molecular Biology* 290: 997–1008)
Image on page 104 (bottom) from the RCSB Protein Data Bank (www.rcsb.org) of PDB ID 1I8F (Mura, C., Cascio, D., Sawaya, M. R., & Eisenberg, D. S. (2001), "The Crystal Structure of a Heptameric Archaeal SM

Index

T

Acknowledgments

I WOULD LIKE TO extend special thanks to . . .

Bryan Todd, Laura and my family, Darrin Wilson, Lorena Ybarra, Dave Seldon, Adam Sugihto, John Paul Mendocha, Matt Gillogly, Jim Runyon, Joel Runyon, Mark Ashton, Andy Martin, Bill Yaccino, Ken Leonard, Bob and Kathy Clapper, Aditya Tiyagi, Josh and Carly Davis, Ivan and Isabel Allum, Wayne Fife, Mark Vuletic, Kelly Huber, Lounis Zenad, Matt Lowry and the Chicago Darwin's Bulldogs, the folks at Infidels, Justin and Crofton Brierley, Hugh Ross, Paul Cook, Wendell Read, Richard Morgan, Doug Angus-Lee, Heather Angus-Lee, Elizabeth Blair York, Marcy Kennedy, Sue Towne, Mark Widawer, Glenn Livingston, Jack Born, Bill and Steve Harrison, Chompasaurus, Ray Glinsky, Johann Chau, Mark Sandford, Paul Braoudakis, Nate and Laura Jennison, John and Jay Fancher, Nathan Beauchamp, Randy Ingermanson, Werner Gitt, Bill Jenkins, Roger Wasson, JD Leman, Tom Hoobyar, Paul Nelson, Bob and Melanie Boldt, Jamie Cleghorn, Heather Treadway, Jen Aldrich, David Deutsch, John Carlton, Gary Bencivenga, Jay Abraham, Jon Benson, TJ Oosterkamp, Charles Martin, Barry Lycka, Jeff V. Cook, Todd Pittner, the late Mike Marshall, Jillian McTeague, and Derek Vasconi.

And a shout-out to the man who first crossed the Berlin Wall without getting shot. Those who followed are forever in your debt.

About the Author

PERRY MARSHALL is an author, speaker, engineer, and world-renowned business consultant in Chicago. With a decade of research, he brings a fresh perspective to the 150-year-old evolution debate. Bill Gates of Microsoft and the founders of Google revolutionized software and the Internet through their status as outsiders. Similarly, this book harnesses a communication engineer's outsider's perspective to reveal a century of unrecognized research and discoveries. *Evolution 2.0* resolves the conflict between Darwin and Design, opening new avenues of science research and raising tantalizing new questions.

Perry's work in digital communications, control systems, acoustics, and e-commerce brings practical insight to questions about nature and science. His books include *80/20 Sales and Marketing*, *Ultimate Guide to Google AdWords*, and *Industrial Ethernet*. He has a degree in Electrical Engineering. He has consulted in over 300 industries, from computer hardware and software to health care and finance.